The Politically Incorrect Guide to U.S. Energy Policy

How the elites have robbed us of prosperity with bad science, bad policies and mind numbing greed

By

Robert N. Henry

Copyright 2015

Robert N. Henry

All Rights Reserved

Dedicated to Dr. Walter Henry,

... a brilliant doctor, researcher and academic ... bringer of echocardiography... and part of the team that made the first voyages to the final frontier.

Sure that was nice ... but being a great brother ... that was even better.

Table of Contents

Preface – Why I Wrote This Book 5

Chapter 1 – The Overriding Principle – Energy is Prosperity 8

Chapter 2 – CO2 and Fossil Fuels are Good Things Not Bad Things 13

Chapter 3 – CRITICAL: Make Thorium the Centerpiece of America's Energy Future 23

Chapter 4 – CRITICAL: Adopt Flex Fuel Vehicles 30

Chapter 5 – CRITICAL: Destroy the OPEC Monopoly 34

Chapter 6 – Encourage Things That Work 37

Chapter 7 – Stop Tilting At Windmills 42

Chapter 8 – Let the Sunshine In 47

Chapter 9 – Bring Me Batteries, More (Better) Batteries 51

Chapter 10 – Making Hay Out of a Lump of Coal 57

Chapter 11 – Creating an American Market for Rare Earth Metals and Thorium 61

Chapter 12 – Bringing Manufacturing Back by Making America the Low Cost Choice for Energy 66

Chapter 13 – Place Restrictions on the EPA 70

Chapter 14 – End the AGW War on Prosperity 73

Chapter 15 – Encourage Homes and Businesses to Drop Off the Grid 81

Chapter 16 – CRITICAL: Strengthen the Electrical Grid and Infrastructure Against Electro Magnetic Pulse Strikes 84

Chapter 17 – The Role of Fusion Power in America's Energy Future 89

Chapter 18 – Conservation- More Energy for Free 92

Chapter 19 – The American Political System, Free Markets and Energy Policy 94

Chapter 20 – SUMMARY: A Master Plan for a Better American Energy Future 99

Chapter 21 – Who Are the True Environmentalists? 102

Chapter 22 – Final Thoughts 105

Appendix - Sources and Recommendations 106

Preface – Why I Wrote This Book

"There are no constraints on the human mind, no walls around the human spirit, no barriers to our progress except those we ourselves erect." **Ronald Reagan**

American energy policy has long been entwined in politics. From oil depletion allowances, to Highway Trust Fund pork barrel spending, to turning our food supplies into fuel, America has been awash in the grimy world of "scratch my back" Washington lobbyists, big money contributions and radical ideologies. The latest to wreak havoc on American energy policies is our old friend global warning. This is a bought and paid for science that leads policymakers to believe that conventional energy is bad and that solar and wind must be their replacements. We have finally reached a place where the politically correct and expedient have triumphed over sound energy policies on a vast scale. This book is a modest attempt to draw attention to the colossal mistakes and missed opportunities that have taken place and to show a way forward to a brighter energy future.

This book is about bad policies, but it is also about new technologies that are on the cusp of transforming America and the world. Some are already in existence, but are not being fully or properly utilized. Others are just around the corner and will need some encouragement to reach their amazing potential. As you go through each of the chapters, try to keep an open mind and ask yourself if the issue discussed has been handled appropriately by our political class. I think the thread that runs though these topics is that we have been short sighted and less than strategic in the policies we have adopted.

While I want to disclose up front that I am a free market libertarian, there is something in this book for people of all political persuasions. If you are on the left, for example, I hope you will give serious consideration to the chapter on Thorium reactors ... it is a form of nuclear energy with virtually no safety issues, little need for waste storage and a zero carbon footprint. It is the perfect compliment for renewable energies (which don't operate on a persistent basis.) It offers the potential for energy cheaper than coal and could offer developing nations a way forward that does not include the proliferation of fossil fuels. If I can persuade the readers of only one thing, it is that this single technology has the potential to unite left and right and provide light, heat, and electricity to an energy starved world.

If there is a second principle that I hope will be embraced, it is the core doctrine that ***energy is prosperity***. So many in the climate change movement wish to end the very thing that has reduced climate related deaths by over 98% in the last 80 years ... namely cheap and abundant energy from fossil fuels. We now live in a world where cold and heat can be transformed to comfortable living by the turning of a dial. We have shrunk the size of the planet by our transformative use of transportation technologies and abundant energies. If a person could be brought here from 300 years ago, they would be amazed at the ease with which we grow food and travel and by the exceptional quality of our air and water. All this has been the result of cheap abundant energy.

The policies advocated in this book will offend the usual list of status quo business's, political hacks and vested interests. It is thus that the name *The Politically Incorrect Guide to American Energy Policies* was chosen. It is my fervent hope that people who truly care about this great country will learn about some of the energy challenges and opportunities in a way they may not have previously considered. A better America and a better world is the goal, but this requires a better way to make decisions. Our current political system favors vested interests over the country's interest. Educating the readers about what is truly best for America is the best way to shine a light on the poor decisions that have been made and to offer hope that someday somehow we will follow a better path.

ABOUT ME

I was raised in a small town in Western Maryland where I somehow managed to set three state records in track and field and get a few Division I scholarship offers for football. One of these took me to Duke University where I studied Economics (with minor studies in Chemistry.) Chance put me on the Duke Debate team and this experience taught me how much I really loved "the art of words and the power of reason."

I have had a lifelong love of science and even though I strayed into a non science career, I always managed to keep up with the latest breakthroughs in science and technology. I have a blog called **newandamazing.com** where I try to offer up equal servings of humor, speculation and wonder about the wonderful world of science. If you want to find out how to build cloud cities on Venus or learn about quantum physics and consciousness or even discover why the sum of all positive numbers is actually minus 1/12, then NAA is the place for you.

Somewhere along the line, I managed to acquire a couple of Masters Degrees in Finance and found myself in a career arranging numbers from tallest to shortest (or something like that.) I even managed a second career writing software, building data warehouses, setting up networks and generally getting electrons to behave properly.

My first love however was always the world of words and so now I am working full time as a writer. I have written three Sci-Fi Fantasy novels and this is my first venture into the world of non-fiction. I decided to write this book because I felt very alienated from science by one topic and that was global warming (or "climate change" if you prefer the tautology infused *I can't be wrong* version.) The complete shutdown of skeptical thought and the seeming rejection of the scientific method annoyed me greatly. As I studied the science behind this world changing theory, I decided it was not me that was outside of science ... it was all those who sold out for a few dollars more, as well as their collectivist masters who arranged the financing.

As I studied the economics of the proposed solution to the CO_2 "problem," I realized that we were dooming ourselves to a cycle of perpetual low growth. Anybody who has ever studied the horrible debt burden that we have taken on as a nation quickly realizes that the only way out of this is rapid growth. The truth was so obvious ... energy brought humanity out of the grinding poverty and limited life spans that plagued our ancestors and only low cost, abundant energy can move us forward to a prosperous future.

The more I studied the fields of energy and politics, the angrier I got. Our treatment of rare earths, flex fuel vehicles, thorium reactors, EMP pulse vulnerabilities and more demonstrated that the political class has behaved very badly. As if the other mistakes were not enough, the political class now seeks to doom us to perpetual debt and economic stagnation by promoting a theory of climate change that is laughably bogus. I had had enough ... so *The Politically Incorrect Guide to American Energy Policies* was born. I hope the book will at least challenge how you look at the intersection of energy and politics. There is a lot of intellectual squalor and deceit at this intersection and changes need to be made.

Chapter 1 – The Overriding Principle – Energy is Prosperity

"Beware of false knowledge; it is more dangerous than ignorance." **George Bernard Shaw**

One of the most important messages of this book is that ***energy is prosperity***. It is an overriding principle that cannot be overstated. If you wish to make America a more prosperous country or if you wish to lift the poor nations of the Earth out of poverty, you simply must find a way to supply more energy. Energy is the very lifeblood of a prosperous society. That is why it is difficult to take seriously those that want to eliminate fossil fuels. Without a replacement that is equally cheap and plentiful, the world will suffer grievously.

Just look at the chart below and you will see that the growth in population and GDP per capita has been coincident with the growth in fossil fuels.

Global Progress, 1 A.D.–2009 A.D. (as indicated by trends in world population, gross domestic product per capita, life expectancy, and carbon dioxide [CO_2] emissions from fossil fuels)

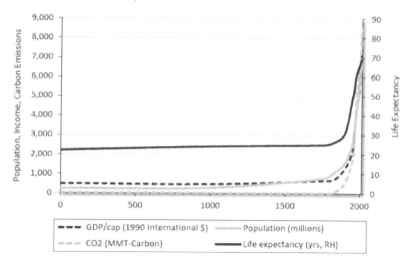

What you see is a true "hockey stick" graph. Population, life expectancy and prosperity (GDP per capita) did not grow until the advent of carbon based fuels. Conversely, it is logical to conclude that any massive cut back in carbon based fuels (without a corresponding rise in other energy sources) would lead to massive increases in poverty and starvation. The view that we simply must end the carbon based economy by such and such a date (usually 2030 or 2050) is not only foolish, it is heartless as well.

ENERGY CONSUMPTION AND GDP

When you compare energy consumption to GDP, you get a fairly straightforward relationship ... if you have the ability to use more energy, you will be more prosperous (GDP per capita). In the chart below, you will see some outliers (particularly among high oil producers,) but in general the top right is where you will find the richer western nations, while the bottom left if where you will find the developing nations.

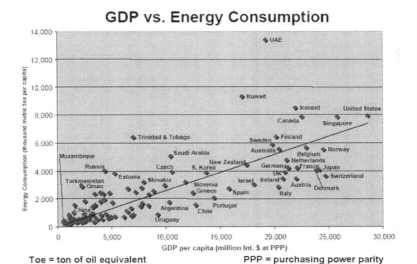

Toe = ton of oil equivalent   **PPP = purchasing power parity**

What does this mean? It means that if you want more GDP per capita, you will need more energy. If you live in a poor country in Africa, for example, your way forward economically is to find a way to get more energy.

THE EFFECT OF PROSPERITY ON POPULATION

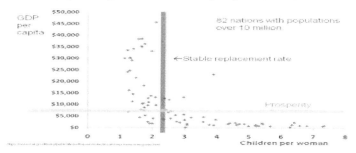

The story gets even more interesting when you consider the effects of population on poverty. (See graph above.) As it turns out, there is an optimum population for growth and prosperity. If your population growth rate is too high, then you have a difficult time growing the economy enough to even maintain the current level of poverty. (People have lots of kids for many reasons, but high infant mortality and the desire to have multiple candidates to take care of them in old age are probably the key reasons.) In any event, if you are in a high birthrate country, you are probably living in poverty. If you are in a stable birthrate country, you are probably living in prosperity.

The question is how to break this cycle? One way to stabilize the population growth rate is to become more prosperous. This is a chicken and egg question, but it is important. If you can introduce enough energy, then you should be able to get on the top /left of the graph. If you do nothing, you will stay on the bottom / right part of the graph. *Energy is the game changer.* Plans that call for redistribution from wealthy nations to poor nations as a sort of climate penance are doomed to failure because the money will invariably go to the strongman of the nation and will not allow for GDP growth (which comes from making things and creating services.) Obviously, it is not *just* about energy (n.b. Venezuela). It is also about adopting a viable economic system and having stable money and incentives for saving, but energy is crucial to wealth. That has been established over and over again (on a global level and on a country by country level.)

HOW THORIUM COULD BE A GAME CHANGER

What we see is that the poorest countries have the highest population growth, which in turn keeps them poor. A country with a super high birth rate that say doubles its population every 15 years needs to double its GDP in 15 years just to maintain the current level of poverty. This is how the rich countries get richer and the poor countries get poorer.

However, as population reaches a certain level of prosperity, they enter a "sweet spot" whereby they begin to slow their population growth and start to achieve real economic growth. And the way poor countries reach the magic growth rate is through market based economics and through finding a reliable source of low cost energy. As you will see in the chapter on Thorium, no energy source offers more promise to transform low income countries than this amazing technology. Thorium reactors offer the promise of "energy cheaper than coal." This could be a game changer because both right and left COULD agree that this technology works for them. For the right, a new source of energy offers the chance for rapid economic growth and energy independence, while those on the left might see this as a zero carbon footprint fuel. Of all the ideas mentioned in this book, this one has the greatest potential for political agreement across a wide spectrum of views.

For the poorest countries, having a technology that can be built on an assembly line and sent anywhere in the world AND offers the potential for low cost energy could be the miracle that transforms their economy.

KEY PRINCIPLES OF ENERGY MAXIMIZATION

Below are a few key principles for wealth creation. They go hand in hand with the core principle of this book that **energy is prosperity**.

1. *Carbon Dioxide is Not a Pollutant* – This is an extremely important principle to grasp. It is covered more thoroughly in Chapter 2, but suffice it to say that CO_2 is the gas of life and is a crucial molecule for all plants on Earth. **The extra CO_2 in the atmosphere is greening the planet and feeding a starving world. It is not causing dangerous global warming**. Just understanding this key principle will allow you to see the world of energy in a completely different light.

2. *It is Insanity to Destroy Our Most Abundant Forms of Energy* – The EPA has adopted limits on carbon emissions from power plants that will destroy the coal industry, much of the natural gas industry and will eliminate from the grid approximately 40% of our electricity capacity by 2020. According to the EPA Director that testified before Congress, this *might* lower worldwide temperatures by .01 degrees (actually the estimate is high.) Seriously, are we really willing to wipe out entire industries, cost tens of thousands of jobs, raise utility rates on the poorest among us and introduce power shortages across the land for the possibility of a .01 degree temperature

reduction? Incredibly does anyone even know what the optimal temperature of Earth should be?

3. *Regulations Should Work With Not Against Energy Maximization* – Congress and the Courts have given the EPA the power to destroy the U.S. economy. Unelected bureaucracies should not have such power. In Chapter 13, the ways in which the EPA can be brought under control are discussed.

4. *Technology is the Key to More Energy* – The pace of change in the modern world is incredible. Energy will also be affected by the changes that sweep though society. New technologies will soon have a profound effect on the world. See Chapter 3 (Thorium), Chapter 9 (Batteries) and Chapter 17 (Fusion) for a few examples.

5. *Growth is the Only Way Out of Our Massive National Debt* – There is only one way to pay off the national debt that threatens to destroy the country and to keep the promises we made on entitlements ... and that is vastly higher growth rates. If we move from 2% to 5%, the difference would be profound. To get to this higher level of growth, we need to have large increases in energy. Efforts to limit energy simply lock us into a future without hope.

Chapter 2 – CO2 and Fossil Fuels are Good Things Not Bad Things

"Whenever you find yourself on the side of the majority, it is time to pause and reflect."

Mark Twain

We frequently hear some pundit or editorial writer talk about "carbon pollution" when describing CO2. Even the Supreme Court in *Massachusetts vs. EPA* (by a 5-4 vote) agreed that the EPA could regulate carbon dioxide, as it was a form of "pollution." This commonly held belief is a spin-off of the great *global warming* scare and is completely false on every level.

CO2 IS NOT A POLLUTANT - IT IS A MIRACLE GAS

While it is true that particulate matter in the air (such as soot) can be destructive, carbon dioxide is a friend of the planet in many ways. Here are just a few:

1. CO2 is a gas used by every living plant. If you give plants more CO2, they will grow larger and produce higher yields. The science behind this has been verified by numerous studies. (See the Appendix for a list of studies done on the effects of CO2 on plant life.)
2. Increased CO2 increases the efficiency of photosynthesis and allows plants to use less water.
3. Because plants need less water when they get more CO2, deserts will shrink as the plants at the margin suddenly become viable. That's right contrary to what we are told, more CO2 actually helps prevent desertification.
4. Plants were evolutionarily designed when CO2 levels were closer to 2000 parts per million (as compared to today's 400 ppm). As CO2 rises, plants are bathed in a world they were meant to be in. (NOTE: 2000 ppm would have NO effect on humans. Indoor air is frequently above that and OSHA limits do not kick in until you get to 5,000 ppm.)
5. Crop yields are "marginally better" to "incredibly better" when CO2 levels are increased. Every greenhouse owner knows this and that is why CO2 is pumped into greenhouses around the world. Yields are always higher and the extra CO2 is not harmful in any way to people.
6. As yields increase, farming density increases, which means there is less pressure to clear cut forests and other natural parts of our world to grow more crops. If you want to save the rain forest and other natural areas, adding CO2 is one way to help.

7. Our world is actually becoming GREENER as CO2 has increased. This can be seen in satellite photos and the actual observed changes from 1982 to 2010 are pretty remarkable.

So what we have is an amazingly beneficial gas that slightly warms the planet and feeds all plant life on Earth. Creating more of it is not to be feared, but celebrated. It is the miracle gas of life and if it goes from 4/100s of 1 percent of the atmosphere to 8/100s of 1 percent of the atmosphere, it is a GREAT BENEFIT to mankind and the natural world.

It is time we claimed the moral high ground in the climate change debate. It is the pro growth side that wants to green the planet, feed the poor, promote freedom, re-establish the preeminence of the scientific method and stop the slaughter of birds from wind turbines. It is the AGW side that wants to impoverish the world by imposing the heavy hand of government.

WHY IS CO2 FEARED?

So why the fear of this gas? We have been bombarded with warnings that the planet will burn if we allow any more CO2 to reach the atmosphere. The truth is that this is virtually impossible. The reason I am able to say this with great confidence is that all forecasts of 3, 4, 7 or 9 degrees of temperature increase rely on positive feedback and not on the heat trapping of CO2. Yes, you heard that right. There is almost no controversy about CO2 only causing a very limited amount of direct warming. CO2 traps heat at various micron levels and lab experiments indicate that if the effect were linear (which it is not) a doubling of CO2 would cause about 1 degree centigrade of warming per century. The dirty little secret is that the IPCC forecasts all rely on secondary warming for which there is virtually no scientific or historical evidence. Even the assumption that a doubling of CO2 would cause a linear temperature increase appears to be wrong. As CO2 increases, there is evidence that the heat trapping effect of CO2 diminishes. (See the link to *Global Warming on Venus* in the appendix at the back of this chapter for an interesting take on the heat trapping saturation limits of CO2).

DANGEROUS POSITIVE FEEDBACK IS INCONSISTENT WITH HISTORY

All the dire forecasts of the IPCC rely primarily on *positive feedback*. As the reasoning goes, CO2 causes temperature to rise slightly and this warming causes more water vapor which causes even more warming. The concept is like the gain in a microphone that suddenly squelches to an annoying level (an actual example of positive feedback).

The problem with positive feedback is that it is rare in nature. If it truly occurred as we go from say 400 ppm of CO2 to 600 ppm of CO2, the earth would have burnt to a crisp a long time ago, as for most of earth's history CO2 levels were vastly higher than today's levels (which are very low from an historical perspective.) Look at the CO2 levels below as plotted against temperature and you will see that there has been almost no correlation between temperature and CO2 on geologic time scales.

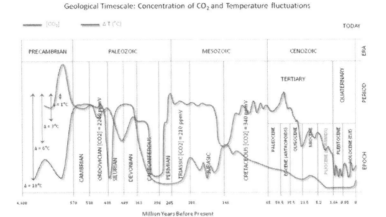

All this history tells us that higher levels of CO2 did not trigger massive runaway positive feedback. It is almost irrefutable that feedback is small or negative. Yet this is at the core of the IPCC models. At one point, we even had an *Ice Age* when CO2 was 20 times higher than today. Notice also, that during the Cambrian and the ages that followed, CO2 was quite high by comparison to today and yet that is when plants emerged and prospered. Plants are truly creatures of a CO2 rich world. Finally, notice how low CO2 is today. We are at about 400 ppm (up from 280 ppm during the recent pre- industrial period.) At 150 ppm, all plants die. Look at the chart again and think about what the real risks are for changes in CO2.

THE TALE OF THE CHICKEN AND THE EGG

While the famous parable of the chicken and the egg may have no answer, the causal relationship between CO_2 and temperature does have an answer and it may surprise you.

Al Gore in his book used a chart that purported to show a relationship between CO_2 and temperature. And guess what, on shorter time scales than geologic scales, he was right ... the two do track together. See below:

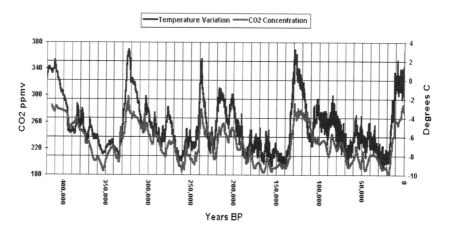

Unfortunately he drew exactly the wrong conclusion from the data ... he got the causality wrong. Studies of ice cores have revealed that temperature comes first and then CO_2 follows. Yes that's right, CO_2 is caused by temperature and does not cause temperature on time scales longer than a thousand years. In fact, there is about an 800 year lag between a change in temperature and the CO_2 in the atmosphere. This is completely logical as gases dissolved in a body of water are released as the temperature of the water is increased and absorbed when the temperature of the water is decreased. This is basic science and it is easy to understand. Whether it is the ocean or a glass of water, this principle holds true. Causality going in the opposite direction has no basic science to support it because there are no proposed mechanisms for past changes in CO_2 other than natural events like volcanoes.

THE COMPLEXITY OF INCREASING WATER VAPOR

As water vapor (a green house gas) is increased, proponents of AGW say that temperatures will rise because of the trapping of heat by the water vapor (positive feedback). This is a very shallow understanding of how the climate system works. The truth is that more water vapor means more clouds, which means more sun reflecting cover above the earth. (see chart below for illustration of "albedo").

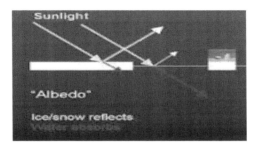

Logically, as the albedo increases, temperatures will drop. Rain patterns, wind patterns and oceanographic movements are all affected by changes in water vapor and it should be apparent to anyone that cloud and moisture patterns are enormously complicated with feedbacks that are both positive and negative. Volcanic activity and solar winds can also affect cloud cover and warming cycles. El Nino cycles, the PDO (Pacific Decadal Oscillation) and other complex factors also are involved. **CO2 as a control knob is a pretty simplistic view of the world.**

Despite the easily derived conclusion that climate is way more complicated than any explanation CO2 can offer, there is another reason to reject positive feedback which can't be overstated. As discussed above, it is Earth's own history. Earth's history should be enough to reveal that runaway positive feedback is simply not possible. This is the GAIA effect at work and it has protected the Earth for hundreds of millions of years.

THE MODELS HAVE BEEN WRONG

If you look at the various climate models that have been put forth, you will see that for the years we can check, the models have simply been wrong. In the graph below, you will see that reality has turned out to be far below the dangerous levels predicted, far below the average levels predicted and even below the lowest forecasted levels of increased temperature. This is because there is NO meaningful positive feedback.

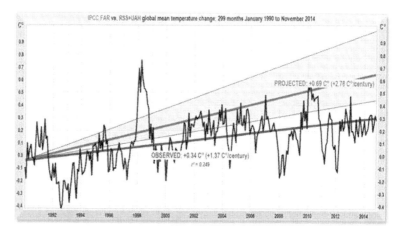

If you believe in the scientific method, then you simply must reject the dire predictions as *wrong*. Richard Feynman, a famous Nobel Prize winning physicist, once said when talking about the scientific method:

"If it disagrees with experiment, it's wrong. In that simple statement is the key to science. It doesn't make any difference how beautiful your guess is, it doesn't matter how smart you are, who made the guess, or what his name is... If it disagrees with experiment, it's wrong. That's all there is to it."

HOW CAN SCIENCE GET THIS SO WRONG

Scientists are paid to do research. That doesn't sound very noble and is a little disquieting to think about, but it is the reality. The money paid by governments of all kinds for climate research amounts to tens of billions of dollars. Unfortunately, they fund only one side of the argument. Governments like any organism want to grow. They see climate change as a way to increase their power over the private sector and as a way to grow revenues. For the United Nations, it is a major opportunity to establish a large scale world government with an equally large tax base. The person who heads the IPCC once said the purpose of the IPCC is "to destroy capitalism." Similar utterances reveal the philosophy that drives the U.N. climate proponents. It is the nature of the beast. Scientists simply follow the money and who can blame them. If I want to study ice core samples and the U.S. government is giving me the money and all I have to do is put a bit of a spin on my research, I will take the money. After a while, I will believe. Tenure and single purpose money are just two of the reasons that true believers are created in academia.

BUT WHAT ABOUT THE 97% OF SCIENTISTS THAT SUPPORT CLIMATE CHANGE?

This argument is repeated by the president, politicians and journalists *ad naseum*. It is completely false. The 97% number comes primarily from a John Cook study at the University of Queensland in Australia. Under this very poorly designed study, a group of researchers looked at over 11,000 peer reviewed papers and assigned their opinion on the topic of AGW to various categories. 6400 expressed no opinion on dangerous manmade global warming. In fact, less than ½ of 1 % of the papers actually said that mankind was causing the majority of global warming and that it was dangerous. Others that said man was causing some warming were thrown into the yes pile. It is a classic case of mixing the apples with the oranges.

Unfortunately, even skeptics would agree with the proposition that man is causing some warming. It is not a question of man causing some warming, it is a question of "Is man causing most of the warming and is it dangerous?" A re-review of the same papers reviewed by Cook et al showed that the number of people who held the view that warming was dangerous and was caused mostly by man, was virtually zero. So we have a study that is repeatedly cited as having 97% support of scientists for the theory of dangerous global warming caused by man which turned out to actually have been stated by only 0.4 % of scientists in the survey. While a significant minority of scientists may hold such a view (follow the money,) the 97% claim is a preposterous claim that should be ridiculed every time it is made. Science is not about a vote. It is about the scientific method.

NATURAL VARIABILTY OR MAN MADE WARMING?

During the 1980s and 1990s the planet did warm. The question is "was this caused by man or was this just a matter of natural variability?" If we can point to an example where temperatures rose for exactly this magnitude and duration (but that occurred before 1950 when CO_2 started to rise,) then it would be very hard to argue that the most recent rise was caused by man and was not simply a case of nature changing the climate in a completely normal manner. You only have to look to the 1930s and 1940s to find just such an example. In fact, the change that took place then was slightly steeper and slightly longer than the change that happened at the end of the 20th century. It simply defies logic to claim that man is causing such variability when it is so easy to find exactly such an example that is 100% natural. Climate changes, it gets warmer and it gets cooler and sometimes it stays the same. The medieval warming period, the little ice age and hundreds of other climate events were all caused by nature. We are very arrogant if we think that we can so easily offset what nature has been doing for millions of years.

WHERE DID MY GLOBAL WARMING GO?

For nearly 20 years there has been no global warming. That's according to satellite records which cover the entire Earth and not just a part of it like the NOAA land records. Satellite records are also less likely to be radically manipulated than land based records. Yes, temperature records are manipulated. It is all very regrettable, but that is the world we live in. Sadly, the most recent temperatures are the most likely to be adjusted up ... this despite the fact that the *urban heat island effect* should cause them to be adjusted down. Machiavelli lives and he lives in the world of temperature data. Trust the satellite records ... ignore everything else.

But ... but ... the heat went into the ocean and is hiding. That seems to be one of the more popular explanations for the GREAT PAUSE. Alas, studies have not been able to back up the claim. The idea that the deep oceans are warmer than ocean surface temperatures just doesn't seem to be consistent with the laws of thermodynamics.

ARE WE AT GREATER RISK OF COLD OR HEAT?

On a very regular basis, the Earth has had *mini* ice ages (actual Ice Ages are even more severe and are on a much larger time scale.) For 400,000 years, the earth has grown cold and then warmed. The cycles have averaged about 100,000 years. The cooling and warming phases average about 90,000 years, while the interglacial warming phases last about 10,000 years. See the chart below:

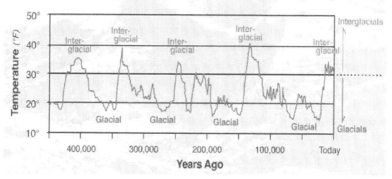

What you will notice is that we are due for another ice age. The Holocene (the name for the current interglacial warm period) has lasted about 11,000 years. This is about 1000 years past the 10,000 year average for such periods. From a big picture standpoint, super cold (with New York under hundreds of feet of ice) is a very real risk. On shorter time scales, we are seeing solar cycles running at extremely low levels. Such levels have in the past been seen during the *Maunder Minimum* and the *Dalton Minimum*. While not actual ice age levels, these periods of cold were quite destructive for man. If only we had something to burn to keep us warm ... wait.

The ability of fossil fuels to warm and cool our environment and to evacuate large populations has resulted in climate related deaths dropping by 98% in the last eighty years. We should keep that in mind as we embark on a path of eliminating fossil fuels. The potential for *Extreme Cold* is far greater than the potential for *Extreme Heat* in the minds of many and we should make sure we can adapt to either by having ample quantities of energy available on a global basis.

CHAPTER 2 - APPENDIX

CO2 SATURATION EFFECT

In my science blog **newandamazing.com**, I try to offer theories that make you wonder about the amazing universe that we live in. One blog post was about *Global Warming on Venus*. It shows that at 1000 millibars of atmospheric pressure (Earth's surface pressure level), the temperature of Venus can be predicted to within 0.5% just by using the distance from the sun, Earth's temperature and some basic laws of physics. With 2400 times as much CO_2 as Earth, Venus tends to validate the theory that there is a saturation limit for the effects of CO_2.

Here is the link if you want to read the entire blog post:

http://newandamazing.shadowsofadistantmoon.com/?p=515

STUDIES SHOWING THE BENEFITS OF CO2

Here are a few studies that show how increasing CO_2 helps to increase plant size and yield:

1. University of Georgia Enviroton Study – Showed substantial increases in yield for both food and non-food crops. One example is that cotton weights at 800 ppm were 6.3 times higher than at normal atmospheric CO_2 levels.
2. U.S. Dept. of Agriculture found that @ 720 ppm with higher ambient temperatures increased yield in sugarcane by 124%
3. At nature.com, a study showed that soy yields were higher in 2002-2006 vs. 1980 levels of CO_2 by a low of 4.3% to a high of 7.6% due to the CO_2 fertilization effect. The key here is that different crops respond in different ways.

For a list of more studies, visit this website:

http://www.co2science.org/index.php (This has actual yield improvements by crop.)

Chapter 3 – CRITICAL: Make Thorium the Centerpiece of America's Energy Future

"In my opinion, there is no aspect of reality beyond the reach of the human mind." **Stephen Hawking**

What if there was a fuel with a zero carbon footprint that was found in such abundance that virtually every country on Earth had lots of this fuel within its borders. What if there was enough of this fuel to last at least 10,000 years? What if energy from this fuel cost less than coal? Who would oppose such a fuel? Well in the 1960s it was opposed by the military industrial complex. Today it is opposed by environmentalists. Both groups were / are extremely short sighted. The fuel, of course, is Thorium and it is the future of energy, not only in America, but around the world.

A DIFFERENT KIND OF NUCLEAR

Thorium is a nuclear fuel. It is this that environmentalists react to. The word *nuclear* seems to be a dog whistle for the green movement and it is opposed at every turn (by most, but not all green organizations.) That is a shame, as it is a completely different kind of nuclear energy. Think for a second about the primary arguments against Uranium Light (or Heavy) Water reactors. These are:

1. They can melt down and release radioactivity into the air.
2. They require lots of water and use steam as the primary heat transfer mechanism.
3. Because the temperature difference between the liquid state of water and the gaseous state of water is so low (100 degrees c), the process requires high pressure (90 atmospheres) to yield reasonable efficiencies.
4. Because of the high pressure, the reactors must be built *in sitsu* and have ten foot think walls to contain any potential leaks.
5. Nuclear fission of Uranium requires a processing cycle involving centrifuges and other steps to separate the U-235 from the U-238 that is found in nature. (Raw uranium is 99.3 % U-238 and only .7% U-235). Reactors used pellets made from U-235.
6. Because the reactor rods create chemicals that stall the reaction, rods must be frequently replaced. This means that the uranium reactor is extremely inefficient. Light Water reactors are 0.5 % efficient. Heavy water reactors are 0.7% efficient. They are both terrible.
7. Being so inefficient, both reactors create waste of approximately 99.5%.
8. Storing the nuclear waste is a big problem. It is the ultimate NIMBY (Not in My Backyard) challenge.

9. The safety process for a uranium fission reactor requires an "active" safety program. Active means you have to really watch what is going on and take decisive action if something goes awry.
10. Uranium reactors produce materials that can be used in making nuclear weapons. It is one of the reasons that western counties are reluctant to give nuclear technology to third world countries.
11. Uranium is rare and expensive and we are running out. In terms of rarity, it is like we are burning platinum for fuel.

Thorium reactors have NONE of these problems. People who lump Uranium and Thorium reactors into the same "nuclear bad" category are simply not thinking clearly about the differences. In fact, there are some environment groups who see the difference and over time it would be expected that opposition will decline.

Let's review the items listed above from the standpoint of a Thorium reactor (typically called LFTR for Liquid Fluoride Thorium Reactor.)

1. They can't melt down because they use passive instead of active safety monitoring. The key here is a freeze plug that allows the reaction to take place. If the electricity goes off, the plug is unfrozen and the material in the reactor simply runs into an overflow tank (with no graphite rods ... thus no criticality is possible.) No one is needed. The safety is simply there by default.
2. Thorium reactors use molten salt for heat transfer. No water is needed. The molten salts (fluoride and other salts) have very high melting points and so the temperature difference for extracting energy is about 650 degrees C. This means you can operate at a single atmospheric pressure and yet still achieve great efficiency.
3. No pressure containment is needed (see above)
4. Since there is no need for a pressure containment unit to surround the reactor, the reactors can be built on assembly lines and shipped any place in the world. Assembly lines mean low cost. The potential to really help emerging nations is a pleasant by-product of this feature.
5. Expensive processing of fuel is not needed. Start-up pellets would need to be manufactured, but after start-up, the feedstock would be Thorium as the isotopic issues that exist with Uranium do not exist with Thorium.
6. Thorium reactors don't have the inefficiency problems that light water reactors have. They are about 99.5% efficient. That makes then about 200 times as efficient as Uranium reactors.
7. Thorium reactors create vastly smaller amounts of waste and many of the waste materials are valuable in the areas of nuclear medicine and alpha decay engines

used in satellites. Waste from Thorium reactors is very small when compared to Uranium reactors.
8. Storage of waste is vastly smaller and can be probably be recycled. Thorium reactors (like travelling wave reactors) can also burn wastes from Uranium reactors, thus serving a valuable social purpose over and above the energy produced.
9. As mentioned, LFTR reactors use passive safety. That is why they pose almost no risk of accidents.
10. Thorium reactors do not create plutonium as a byproduct and thus it is very difficult to create nuclear bomb material from these kinds of reactors, (That is why liquid fuel reactors from the 1960's were rejected by the military industrial complex.) See the appendix to this chapter to learn more about the history of the two types of competing reactors and the great men who championed each.
11. While Uranium is rare and getting harder to find, Thorium is common. It is in fact as common as lead. There is enough Thorium to last at least 10,000 years and probably a lot longer.
12. BONUS ADVANTAGE FOR THORIUM – LFTR reactors can be used in space colonies as they require no water. So whether it is colonies on the moon or underground colonies on Mars or even cloud cities on Venus, Thorium reactors will almost invariably be the energy generator of choice.

WHAT IS THORIUM?

Thorium is an element with an atomic number of 90 and it was discovered in 1905 and named after the Norse God Thor. It has a very long half life (about 10 million years).

Half lives for nuclear materials are counter intuitive. You would think that if something has a long half life, it would be very radioactive. Just the opposite is true. You could hold a ball of Thorium the size of a golf ball in your hand and you would be perfectly fine. Don't be fooled however, as that ball (or one just a bit bigger) contains enough energy to meet the needs of your entire lifetime. This is energy density on a large scale.

The actinide metals (atomic numbers from 89 to 103) are made in stars that go supernova. Earth wound up with Thorium and Uranium and other such elements in its core because of their weight. Volcanic and other geologic processes have deposited some Thorium on the surface. Thorium and Uranium are one of the reasons the Earth's core remains hot (from the alpha decay of these very special elements.) Note: Alpha decay is what happens as radioactive materials decay. As they decay, they give off heat.

HOW DO THORIUM REACTORS WORK?

It is ironic that Thorium reactors are really Uranium reactors (in that Thorium decays to U233 inside the reactor.) U233 is one of the isotopic forms of uranium not found in nature. If you look at the reactor diagram below, you will see all the key parts. There is the liquid fluoride which is used as a heat pump to extract energy. There is the freeze plug which is there to provide passive safety. There are fuel stock components and there are a number of waste extraction mechanisms to consider.

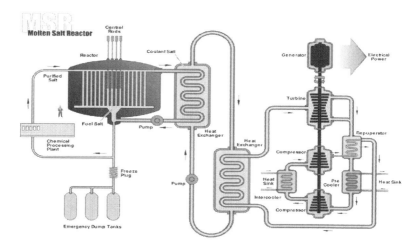

Like all nuclear reactors, LFTR reactors are complex and pose a number of technical problems. Chief among these is the corrosive nature of the liquid fluorides. This poses a materials challenge that will have to be solved. There are also the regulatory challenges as well, as Thorium is now considered a waste material that must be disposed of when mining rare earth elements. (See Chapter 7 for a discussion of this topic.) In China, they see Thorium for what it is and great stocks of it are being saved for future use.

IF THORIUM IS SO GREAT, WHY DON"T WE USE THIS TODAY?

The U.S. actually built a liquid fuel reactor in the 1960's and 1970's. The same person, Alvin Weinberg, who developed the light water reactor in the Oak Ridge, Tennessee research labs, also designed and built a Thorium reactor as well.

Unfortunately, many powers were arraigned against him as the military dominated in this arena. (See the Appendix for more on this history). When the decision finally came down in 1971 to shutter the liquid fuel reactor program, a great opportunity was lost. While nuclear energy has been the safest form of energy in the world, fears of have it have caused costs to soar and new plants to simply die on the drawing room floor.

So why no Thorium energy today? Thorium reactors have gone completely unsupported and the existing powers that be don't want to upset the "sell the fuel pellets" model for existing uranium reactors. I mean money is given to the political class and viola no LFTR. This is a core principle of politics as new technologies that are disruptive and require government support in terms of regulation etc. simply can't overcome regulatory and political inertia. Who really is for the technology ... a few geeks perhaps, but certainly not the nuclear industry or the petroleum industry (a competitor)? With the public having almost no knowledge of the technology, there is no political momentum. In fact, the political momentum is all on the other side. There is a way to break this logjam, but it will require a leader with vision.

HOW TO MAKE A BREAKTHROUGH WITH THORIUM

One of the most effective ways to make something happen in the area of technology is to offer a prize. Think of the Millennium Awards for outstanding math challenges (one has already been solved) or the X-prize for putting private rockets in space (SpaceShipOne won the award.) Sometimes only a modest award is needed to jump start a breakthrough.

One thing that would start Thorium on its way is to offer an X prize for the first company to build a Thorium reactor that can be built on an assembly line and shipped to a new location. A prize of 3 billion dollars should generate a lot of interest. The prize should be only for American companies and it should include R&D tax credits for any company willing to try to win the prize. For a grand total of say $7 billion in prizes and tax credits, you would probably have a working thorium reactor within 3-5 years.

THORIUM AS A TRANSPORTATION FUEL

If you have cheap electricity, you have the potential for transportation fuels. One such fuel would be hydrogen as electrolysis would be a straightforward way to create something to burn. Better fuels like artificial gasoline (Synfuels) could be created by combining the heat of a LFTR operation with hydrogen (easily obtained from water at high temperatures) and carbon (extracted from the air if you are CO_2 phobic or extracted from coal or even from smokestacks that contain CO_2.) Synfuels would require no drilling for oil and could be easily manufactured given a heat/electricity generating LFTR operation. Even methanol would be fairly easy to produce if super cheap electricity were available. Just because a power source is stationary, doesn't mean it can't create fuels for use in vehicles that move. Heck, electric vehicles with super cheap nano batteries (coming) might allow direct power for cars from LFTR generated electricity.

FINAL THOUGHTS

Thorium is the most promising technology I encountered. There are vast quantities of it and small (in terms of our national budget) sums of prize money might make this happen rapidly. Getting rid of the regulatory burdens in this area are also necessary to allow this technology to emerge. When it does emerge, it will be a game changer. It is also one of the few technologies that has the potential to unite right (energy independence, rapid growth, destroy OPEC) and left (no CO_2). It will take a united front because both BIG OIL and BIG NUCLEAR will fight this. It is a fight worth having however.

CHAPTER 3 – APPENDIX

The story of how all nuclear reactors were basically designed by the Navy who wanted something that would fit in a German U-boat Model 27 (later to become the Polaris) is a pretty incredible story. It pitted Alvin Weinberg on one side against Admiral Rickover on the other. Light Water reactors won. Thorium lost. A good summary of this story can be found on my newandamazing.com website. Here is the link:

http://newandamazing.shadowsofadistantmoon.com/?p=465

Chapter 4 – CRITICAL: Adopt Flex Fuel Vehicles

"A government that robs Peter to pay Paul can always depend on the support of Paul."

George Bernard Shaw

Flex Fuel Vehicles could have a tremendous impact on the transportation market in America. They would allow us to burn natural gas (in the form of methanol) to move people and goods from place to place. It would be a powerful step in truly breaking OPEC and it would radically reshape the country's need to run our cars on oil. Unfortunately, the U.S. has all but outlawed flex fuel vehicles. What follows is the story of a bought and paid for energy policy that caters to the needs of BIG OIL and BIG GOVERNMENT. It is a fascinating tale.

METHANAL - THE AMAZING GASOLINE SUBSTITUTE

What if you could go to the pump and instead of seeing only three choices for powering your car, you could see more? (NOTE: The choice between regular, medium grade and high test is not really a choice. It is just oil with a slightly different pedigree.) Picture what it would be like if the additional choice or choices were fuels that were 100% made in America and lower in cost per gallon. The main product I am talking about is methanol and it is amazing. Methanol can be made in any number of ways, but it is easily made from natural gas which the U.S. has in mega abundance. So imagine you are looking at the pump and you see that gasoline is $3.50 per gallon and you see that this thing called methanol is selling for $1.29 per gallon. Which would you choose?

Oh you say to yourself, if only I could actually run methanol, this would be my most fortunate day. Alas, my car isn't set up to run methanol or ethanol (except in certain gasoline mixes) so sigh, I guess I will just have to pay the $3.50. You think it would probably cost thousands to obtain a car that could run on methanol so you just keep doing what you have always done. "Fill 'er up."

This picture has one major flaw ... it is illegal to sell methanol at the pump in every state. Yes selling a fuel that costs less and comes only from America is illegal to sell. Think about that.

SO WHAT WOULD IT COST TO PRODUCE A CAR THAT CAN RUN METHANOL?

The vehicles that are capable of running gasoline, methanol and ethanol in any blend are called FLEX FUEL Vehicles. The cost to turn any car into a flex fuel vehicle is approximately $90. Yep that's it. For a fee so small it would almost be unnoticeable at the lot, you could have a flex fuel vehicle that could run almost anything. If you have a used car, the cost would be more depending on year and model, but it would still be quite low. Under today's law, converting your car to something really useful like a flex fuel vehicle would be illegal.

HOW DO FLEX FUEL VEHICLES ACTUALLY WORK

The really cool thing about today's modern engine is that they sample the fuel many times per second so that the carburetor settings are always optimal. A flex fuel vehicle would allow you to put any combination of gasoline, methanol and ethanol into your tank. There is no need for separate tanks, just make your choice at the pump and you are all set. The car will figure out how to run the particular mix that you have. Nothing could really be more simple.

BOUGHT AND PAID FOR POLITICIANS

So we have something that would allow true energy independence, saves money for everyday consumers and would break the power of OPEC and it is illegal at every step ... from manufacturing to auto conversions to fuel sales ... such a country.

How is this possible you ask? As with most questions, you must start by following the money. Who doesn't want competition for their gasoline products? Who likes things just the way they are? BIG OIL and OPEC come immediately to mind, but how are they able to control products, fuels and services with policies that work against the American people. Why is it that the champions of the status quo are politicians? Yes that right ... contributions. With political contributions placed in the right hands, "interested parties" can make sure that their products are kept safely away from the nasty brutish competition that would be made possible by freedom of choice at the pump. Who really is on the other side? Well almost no one. If you don't know what is possible, you won't be able to be an advocate. That is how our politically corrupt system works.

WHAT WOULD TRUE FREEDOM OF CHOICE AT THE PUMP MEAN?

If you had true freedom of choice at the pump, you would buy the product that is the cheapest or perhaps the product that works best for you on any given day. For example, since methanol and ethanol have lower energy densities than gasoline, if you were going on a long trip you might decide to fill up your car with 100% gasoline. This would allow you to go further on a tank full of fuel. On the other hand, if you are just driving around town, you would probably just select the lowest cost fuel per mile.

To allow you the proper choice, the *gasoline cost per gallon equivalent* would be shown for all fuels right at the pump. That would make it super easy to save money if the prices got a bit out of line.

But the truth is, they would not get much out of line. Competition would keep all fuels within a tight cost range. This means that OPEC could not set the price at the pump as they do now. Instead the price would be set by the cost of natural gas. This would crush OPEC and make for a far more competitive energy environment for consumers.

TAXES - A DIFFERENT KIND OF PRICE MANIPULATION

Unfortunately, the Empire would probably strike back by raising the level of taxes that would be set for each product. Currently state and federal taxes for gasoline far exceed oil company profits. If you introduced a competitor product that really lowered prices, you would undoubtedly see political pressure to use taxes to hurt the new competitor. This should be resisted at every turn. The benefits to the American people are just too great

HOW DO WE KNOW THAT FLEX FUEL VEHICLES REALLY WORK?

We know these work because we make them for export. Yes, the auto companies sell such vehicles in Brazil and elsewhere. Brazil made a real commitment to get off their oil addiction a few years ago and now they make their fuels from agricultural products. In general, using food products or food acreage to make fuel is not such a great idea (because it raises the cost of food) but this works for them. The experience of Brazil has proven that flex fuel vehicles work and are cheap to convert.

WHAT NEEDS TO BE DONE

What needs to be done is really quite simple.

1. Congress needs to allow/require car companies to build and sell flex fuel vehicles for vehicles of all sizes. Currently there some flex fuel vehicles built in limited quantities but without pump availability, these are of little value. NOTE: While it is generally better to allow the market to work, it seems like a date certain for all gas vehicles to be flex fuel vehicles would be optimally beneficial. With a cost that is so low, it is a small burden to place on auto manufacturers to transform the way we power our cars.
2. Methanol and Ethanol should be allowed to be sold at the pump. All federal restrictions and regulations should be lifted. This can be easily embodied in a single piece of legislation.
3. All restrictions on the conversion of existing vehicles should be lifted. The market place and the consumer should decide if this is something that they want to do.
4. States should pass a Uniform Flex Fuel vehicle law so that all restrictions at the state level would be lifted. It does no good to pass federal laws allowing flex fuel vehicles if the states can just block the technology at their level.
5. Taxes at both the federal and state level should be on a *gallon equivalent basis* so that no product (gasoline, methanol or ethanol) can be treated unfairly.
6. Subsidies for ethanol should be stopped. If it makes sense to grow fuel, so be it. But we should not be subsidizing food discontinuities for fuel reasons when we have so much natural gas/ methanol.

FINAL THOUGHTS

This is just a quick overview of the problem that we face getting the right energy source to the right market. The "oil only" laws no longer serve the transportation sector or the American people and it is time we made the change that Brazil has already made. We are mega producers of natural gas. Let's make the oil from OPEC compete with our strength and not just allow them to dictate the market.

For a great read on this topic, please see *Petropoly* by Anne Korin and Gal Luft. They call methanol "the Rearden metal of fuels." (This is of course a reference to *Atlas Shrugged*.) You can also visit fuel freedom.org to learn more about the flex fuel vehicle story. While their take on flex fuels as a way to reduce global warming is very different from my own, they are real allies in getting Congress and the states to do the right thing. Just as with Thorium, there should be goal congruency between those that want to *save the planet* and those that want to see a prosperous America.

Chapter 5 – CRITICAL: Destroy the OPEC Monopoly

"My big focus is China and OPEC and all of these countries that are just absolutely destroying the United States." **Donald Trump**

OPEC is not a friend of America. They are instead an organization dedicated to extracting the maximum possible cash from the rest of the world ... mainly the West. Saudi Arabia can produce a barrel of oil for LESS THAN $2.00. They have often sold it to us for OVER $100 /barrel. So they pocket huge sums of cash. And what do they do with this cash? They purchase domestic tranquility by giving tens of billions to Wahhabi religious organizations and by offering generous payments to the military and the masses. They also aggressively export their particular branch of Islam by supporting mosques around the world (including America.) Fifteen of the nineteen 9-11 terrorists who crashed planes into some of the nation's most iconic buildings came from Saudi Arabia.

But OPEC is not just Saudi Arabia. Iran, who now has a "get out of jail" card with the new nuclear treaty, will soon be exporting huge quantities of oil. Like Saudi Arabia, these extra funds will go to causes not favorable to America. The chants of "Death to America" still echo through their streets.

WE ARE FUNDING OUR ENEMIES

Think about where Western petro dollars go. They go to Saudi Arabia, Iran, the Arab Emirates, Iraq, Russia, and Venezuela to name a few. This is a veritable enemies list for America. Dollars sent to these places are transformed into terror, mischief and trouble on a global scale. An American political candidate was once asked what he would do about Iran. He said he would lower the price of oil. Eyes rolled, but the answer was a very clever one. Lowering the price of oil takes money directly from the pockets of our enemies and this should be a core principle moving forward. Destroying OPEC should be a key component of any foreign policy.

BUT HOW TO DESTROY OPEC

A core principle is that *energy is prosperity*. The country and the world need lots of energy to create robust economic growth. But this does not mean we should pay any price for energy. A basic principle of economics is that high prices increase the attractiveness of alternatives. So it should be for oil. As discussed in the section on flex fuels, the U.S. can easily switch our transportation fuels from oil to natural gas. We are importing oil. We have lots and lots of natural gas. So simply by adopting flex fuel vehicles and changing the restrictions on selling methanol, we create an alternative product. This would put enormous pressures on oil based products and hurt OPEC.

In addition to alternative products (including Thorium in the long run), you can rearrange the volume use of petroleum by the use of taxes and incentives. Taxing imported oil and using the money to stabilize prices and/or engage in energy research is more or less the exact opposite of what OPEC has been doing to us. When U.S. technology starts to deliverable abundant energy, OPEC has used low prices to temporarily disrupt domestic production and the energy capital markets. They introduced instability to win in the long run. By taxing foreign oil outside of the North American Free Trade Zone (to be established), we can introduce stability to our markets and instability to OPEC. In effect, a properly applied tax would take part of the vast margins for OPEC based oil and convert it to American dollars. If they refuse to sell under such an arrangement ... so much the better. It just makes America an energy independent zone and it reduces demand for their products on a worldwide basis – both worthwhile objectives.

THE DOWNSIDE TO DESTROYING OPEC

The biggest risk to winning an economic war against OPEC would be that they would probably no longer support a dollar denominated currency for the sale of their products. The Chinese and the Russian have long wanted to strip America of its special status as the world's reserve currency. If it were not for our massive deficit spending, loss of reserve currency status would not be much of a risk. A rule in any war with OPEC is that America needs to operate on a balanced budget. If it is not prepared to live within its means, then only a skirmish with OPEC will be possible. However, with better energy, tax, budgetary and regulatory policies, the U.S. growth rate would be substantially higher. This is the only feasible way to stop the rise in the national debt and allow us to live within our means. Ironically, the current administration's policy of limiting U.S. energy production and maximizing taxes, regulations and debt will lead to an even greater dependence on OPEC.

WHICH WORLD DO YOU WANT TO LIVE IN?

We have for decades lived in a world dominated by OPEC. Trillions of dollars have been drained from our economy and sent to people who hate us. Is this not war "by other means." Do you want to continue to live in a world of limited growth, high unemployment and declining prosperity? Do you want to pay for this managed decline by mortgaging America's future ... putting burdens on our children and grandchildren from which they will never be able to recover? Do you want to add to this decline by cutting off energy production in the name of the "God of Warming" when we know that *"Energy is Prosperity?"* Do you want to give up the freedoms that were central to the founding of America, so that power can be consolidated with the political class and mega industries /cartels? Most people do not want to live in this world. It is bleak ... unwelcoming. It is a world we should reject.

Instead, there is the world of freedom and growth that in the past has defined America. It is a "shining city on a hill." It is there, waiting for us, if we have the courage to reject the policies of failure. We are blessed to have vast stores of energy and an energetic and innovative population. We also have the capital (both financial and intellectual) to build a thriving future. Why would we allow self serving politicians and their "clients" to limit what we can be?

Chapter 6 – Encourage Things That Work

"There is no darkness but ignorance." **William Shakespeare**

U.S. energy policy has many things that work. These should be supported. The amazing contributions of the private sector have brought about a revolution and made the U.S. the number one producer of oil and natural gas in the world. We are blessed to not only have the resources that can be utilized, but we are blessed with human capital, technology and capital flows that have brought the core of our prosperity from deep beneath the ground to the surface of our world. Energy is prosperity and we should always be grateful for the resources that we have.

FRACTING AND HORIZONTAL DRILLING HAVE BEEN MIRACLES

The combined new technologies of hydraulic fracturing and horizontal drilling have made an enormous difference in America's energy production and we should do all we can to encourage these technologies. It is ironic that the U.S. (who didn't sign the Kyoto protocol calling for a reduction of CO_2 emissions to 1992 levels) is the one major Western country that came closest to meeting the requirements for CO_2 reduction. The European CO_2 fear mongers are still trying to figure out the optimum way they can destroy their economies so that they can meet arbitrary CO_2 reduction schedules. The U.S. did what the Europeans could not do because we used fracking, et al to produce huge quantities of natural gas which began to replace coal in power generation. This switch to natural gas not only lowered prices, but it lowered CO_2 emissions (not necessarily a good thing.) Fracking is of course banned in most of Europe (thanks to anti-fracking funding by Russia who wants Europe to be dependent on Gazprom (the Russian gas delivery company) to keep the silly Euros warm in the winter.

Just like Europe needs a fracking revolution, many states need to realize that they are punishing their own populations by clinging to anti-fracking laws. New York, Maryland and others have limited their access to low cost energy and thus to rising prosperity. They should change their laws as should European countries that want to prosper.

NOTE: Fracking seems to be undergoing a Moore's law of sorts. According to a story in *Breitbart News*, the productivity of fracking created wells has been doubling about every 2 years. This is just another incredible feature of this innovative technology.

PUBLIC LANDS STILL OFF LIMITS

One of the major failures of U.S. energy policy has been the limits that have been placed on drilling on federal lands. All of the incredible quantities of energy that have been produced with new technologies have been done on private or state owned land. The federal government which owns nearly 50 % of the lands west of the Mississippi has been extremely short sighted in its energy policies. See map.

The Feds have set strict limits on drilling because the environmental movement has pushed so hard to shut down fossil fuels. The near religious belief in AGW has caused a virtual complete blackout of drilling on federal lands. In Alaska, drilling prohibitions in the Arctic National Wildlife Refuge (ANWR) has blocked 10.4 billion barrels of recoverable oil. Thus, we are deprived of 800 million barrels of oil per year. With the use of horizontal drilling, only tiny sections of ANWR would be affected if oil drilling were allowed. The environmental impact would be essentially zero ... yet no permits are granted. What the state wants is also of no consequence. As millions of barrels of oil go un-pumped every day, one wonders what ever happened to the 10th amendment.

THE KEYSTONE PIPELINE

Energy is not only prosperity, energy is jobs. By delaying the keystone pipeline for literally years, the administration has cost thousands of American jobs. In addition, the safety of transporting the oil has been compromised because much of the oil now goes by railcar. Railcars are owned by major administration donors so they are used. But the accident rates are far greater than they are for pipelines. We sacrifice safety and jobs so that *pay to play* can be protected. Even the belief that CO2 will be saved is clearly a myth. Does anyone really believe that the oil from Canada won't be sold to somebody? This whole issue is nothing but corrupt politics at its worst.

THE ALASKAN PIPELINES

Bringing energy from Alaska to the lower 48 would be tremendously valuable to the U.S. economy. Naturally it is opposed at every turn. The environmental concerns about pipelines in Alaska have been shown to be overblown as the pipelines that have been built have turned into migratory pathways for Caribou and various other animals. Apparently they like the heat. The key here is that power lies with the Federal Government and not the State. As measured by acreage, the feds own 61.8% of Alaska. This one fact means that nothing is going to happen without approval from the EPA and other federal agencies that are anti-growth. The one plus of the current stalemate is that we are preserving reserves for future use ... at least that is the hope.

NUCLEAR POWER WORKS

The accident rate per megawatt of electricity for nuclear is virtually non-existent. Coal, hydro, solar, wind, natural gas and oil, and all other forms of generating electricity have meaningful death rates associated with their extraction and use. Nuclear does not. The problem with nuclear is not the actual dangers, it is the irrational fears that accompany all things radioactive. Nobody died at Three Mile Island despite the sensational movies and general hysteria that surrounded the event. Nobody got sick. This was the worst of all U.S. nuclear accidents ... and nothing happened. The cost of nuclear is much higher than it should be because we have mega quantities of green litigation. You want to build a new plant ... good luck.

The way forward for nuclear is Thorium reactors. It is vastly more efficient than traditional light water reactors and it has none of the safety problems of typical reactors. It can't melt down, doesn't need a multi-atmosphere containment dome, it doesn't use water, and uses a fuel that you could hold in your hand. It even burns spent fuel rods from uranium reactors. This is the fuel that the green movement should be behind. Maybe in time they will be, but for now anything that has the word reactor in its name is anathema to the environmental movement. It's a shame really, because this is the zero carbon footprint fuel that makes sense, The two sources of energy they do endorse ... solar and wind ... range from marginally useful to terrible (see chapter on wind power.)

HYDRO POWER WORKS

About 5% of US energy comes from hydro-electric. This is about the percentage limit for this kind of power, but it is very valuable. Water flows day or night and so you have a form of energy that is always available, has a low carbon footprint and is cheaper and more reliable than sun or wind (which suffer from the intermittency problem.) Naturally hydro power is opposed by the green movement.

BIO FUELS ARE MARGINAL

While I love the innovation that has gone into switch grass fuels and algae fuels and all the other forms of bio-fuels, I always have the question how scalable are these? As a fuel at the margin, yes they might contribute to the energy picture in some way, BUT as a large scale project, replacing coal/ natural gas/ or even hydro, I just don't see it. They would be replacing agricultural lands at scale and that would be a repeat of the ethanol problem ... rising food costs, no real CO_2 savings (OK by me, but it defeats the stated purpose) and very little if any net energy created. Until there is a real breakthrough, I will remain a skeptic concerning the value of this energy source.

OTHER FUELS

There are so many ideas for fuels that it is difficult to keep track of them. Basically if it burns, it's a candidate. Of all the ideas that futurologists might tout, only one seems promising on a decadal scale and that is nanotechnology. The ability to create energy from entities that mimic nature (see Chapter 8 for more) is really exciting. As nano tech matures, the potential for breakthroughs in artificial limbs and bodies, alternative immune systems, energy storage systems and energy creating bio-mimicry trees and plants is just off the charts exciting. It is one of those technologies that may change our world in ways we can't imagine. What we do know is that the pace of change in the 21st century will be incredible and that energy breakthroughs will lead the way.

Chapter 7 – Stop Tilting At Windmills

"If you torture the data long enough, it will confess." **Ronald H. Coase**

Windmills are the darlings of the green energy folks. Let's all wander through the giant windmills with our tie-dyed shirts and a flower in our hair and enjoy all the free energy that comes from Mother Nature's own fresh air moving across the plains. Why it's a zero carbon form of energy that will allow us to save the planet ... NOT. Everything about windmills is wrong. Instead of subsidies, anyone who tries to build one of these things should be shot on sight (OK don't do that.) Let's just review some of the reasons why.

THE MYTH OF THE ZERO CARBON FOOTPRINT

It takes about 3 years to recover the carbon energy used to make and install a solar panel. It takes about 18 years to recover the carbon energy used to make and install the giant steel torture devices we call windmills. The windmills themselves have a lifespan of about 20 years. They barely breakeven on the carbon saved. BUT wait ... over time windmills become less efficient so the 18 year carbon payback is actually closer to 22 years. That's right "carbon free" windmills actually use more carbon to create and install than they save during their lifetime. A better way to "save the planet" would be to set a few barrels of oil on fire or perhaps start burning old tires for fuel. In terms of the amount of carbon saved (which is less than zero) these would be equivalent ideas.

WINDMILLS ARE UGLY

Windmills are the ultimate in NIMBY (Not In My Back Yard.) The reason is that they are eyesores upon the landscape. Whether they are ruining a coastal view or ruining the view from the fruited plains, windmills are unattractive. If only this were the worst of it.

WINDMILLS MAKE YOU SICK

People who live near windmills report that the constant tearing at the wind makes them sick. Yes you can always move away, but as a general rule when something makes you sick, you should ask questions about whether that thing is really safe. This is a minor factor to be sure, but it is just another reason why windmills are undesirable. This issue is also one of the reasons why land around windmills depreciates so rapidly.

WINDMILLS SAVE EVEN LESS THAN PLANNED BECAUSE OF INTERMITTENCY

The wind doesn't blow all the time. This simple reality means that you need full back-up capacity from a source of energy that is scalable and immediate. So that means that the coal used to produce a given amount of electricity is virtually the same with or without windmills. In Germany, they decided to go green by making a major commitment to use wind power. The use of coal since 2000 has fallen by ... wait for it ... 7%. Because there are times when the wind doesn't blow at all, you need full backup capacity. That means no savings on capital expenditures for the "bad" kind of energy ... just some variable savings intermittently. The intermittency problem coupled with the transportation problem (see next) makes wind power next to useless.

WINDMILLS HAVE HIGH TRANSMISSION COSTS

Tautology warning: You have to put windmills where the wind blows. This is usually not in cities or next to factories. You put them way the hell out in the hinterlands or you put them way the hell out to sea. That's where the wind is. The problem is the power needs to be somewhere else. It is the cost of getting energy from where it is generated to where it can be used that adds to the cost of windmills and makes them less economically viable than they might appear to be on the surface (which isn't very viable without subsidies.)

WINDMILLS KILLS BIRDS BY THE MILLIONS

Each year windmills chop and grind about 10 million birds. These devices may kill quickly, but they sometimes kill slowly by maiming their prey. The lack of outrage by environmentalists is really appalling. Oil pipelines create warm migratory pathways for animals to take and yet these are "bad." Windmills kill and maim by the millions, yet these are "good." Not only do they kill and maim ordinary birds, but they kill some of the most beautiful creatures on the planet such as hawks and eagles. Here is what the environmentalists endorse.

It is worth noting that the windmill lobby makes claims about deaths of birds and bats that are very low ... 555,000 birds and 880,000 bats per year. Based on studies done in Europe with similar volumes, their estimates are off by a factor of 10-20. Worse, there is even fear that some bird species could go extinct as a result of the slaughter (including the noble golden eagle.) This is really a tragedy without any benefit.

NOTE: The windmill lobby claims bird deaths are OK because cats kill a lot more birds (true.) But I ask you ... how many cats really hunt bald eagles or golden eagles? The large raptors are where the environmental disaster is really taking place.

SUDIDIES FOR WINDMILLS MAKE US POOR

Each year we spend about 12 billion dollars (thehill.com 2-24-15) on subsidies for windmills (federal only). Yes the very same windmills that are not economically viable without subsidies and that save no carbon whatsoever, we pay people to build. We give them money so that they can kill birds by the millions and laugh all the way to the bank. Some estimates put federal and state subsidies for some projects at up to 80% of the cost of building a windmill (massmegawatts.com)

A much better idea would be for each American to go to their local bank and withdraw 100 dollars from their account and then set it on fire. Greenies might not see windmills as the functionally equivalent of burning money, but an economist would see the two activities as being functionally identical on a macroeconomic level.

FINAL THOUGHTS

Windmills make environmentalists happy. I have no idea why. If there is any group of people who should be horrified by the mass slaughter of birds (including some of the most spectacular raptors in the world) it should be them. But no, the fear of global warming is so deeply ingrained, that anything that can make them feel noble and pure about getting rid of fossil fuels needs to be subsidized (even if it doesn't really work in saving CO_2). While stopping the subsidies for these avian torture devices will have an impact, as windmills produce about 1.6% of our energy needs (4% of electricity which is 40% of the total,) it would start to reclaim the moral high ground for a sound and logical energy policy. Ending these monstrosities is above all a moral choice and people of every political stripe should support such a decision.

A reasonable proposal would simply be to stop all wind subsidies (honoring existing contracts of course.) This would remove them from the landscape over time. The opposition would be fierce however, as there is now a powerful wind lobby who "gives" money to the political class. It is a model we see again and again ... get a subsidy ... give money to politicians ... grow the subsidy.

The other model that windmills represent is "confusion is good." The subsidies for wind are now so complex and come from so many different sources, that doing any real cost analysis is virtually impossible. Windmills claim they can create energy cheaper than coal. With subsidies approaching 80%, who is to say they are wrong. My plan is simple ... stop ALL subsidies ... wait 10 years ... see if any are still running. I can't wait.

Chapter 7 – Appendix

Here are a few articles about wind not saving CO2:

- http://joannenova.com.au/2011/07/lessons-in-wasting-money-use-more-wind-and-solar-and-emit-just-as-much-co2/
- http://hockeyschtick.blogspot.com/2013/12/new-report-shows-wind-power-doesnt.html
- http://www.forbes.com/2011/07/19/wind-energy-carbon.html

NOTE: The wind industry claims that carbon payback is just a matter of a few years. They fail to take into consideration the intermittency issue as well as a number of other net negative factors. Be careful when you see extremely short CO2 paybacks being touted by the wind industry.

Chapter 8 – Let the Sunshine In

"I would like to die on Mars ... just not on impact."
Elon Musk - godfather of the solar electric/storage system

Solar subsidies are problematic on two fronts. First they are anti free market. They distort energy costs in ways that change how capital is allocated. Secondly, they support technology that is still not optimally efficient. Nonetheless supporting solar is a net positive development and the U.S. should continue to support solar at the consumer level for the reasons mentioned below.

WE NEED CONSUMER TAX CREDITS NOT CORPORATE WELFARE

Unfortunately, huge quantities of money have been spent on corporations engaged in solar manufacturing. These have been for the most part political payback for support of the political elites. Subsidies granted to Solyndra and several other companies that made political contributions to the current administration appear to be naked *pay for play* practices. Alas, this is not an uncommon model among the political class in Washington. Money granted to these companies was simply thrown away (over $800 million in the case of Solyndra.) The government should be prohibited from being in the investment banking business. The temptation to engage in *pay for play* is simply too great.

Consumer subsidies, on the other hand, have created demand for solar installations that have boasted volumes and made for economies of scale. Over time, these price supports have allowed the price of solar to drop significantly. Eventually, they may even become cost effective on a no subsidy, full cost basis. But cost alone is not the main reason why solar panels are attractive.

SOLAR CELLS AND GRID INDEPENDENCE

One of the great risks that the nation faces is grid destruction by way of asymmetric warfare (EMP weapons) and by way of solar flares. See the chapter on *Strengthening the Electrical Grid Against Electro Magnetic Pulse Strikes* for more details. Solar cells have the potential to be encased in grounded faraday cages which would allow them to survive an EMP attack. The country is so vulnerable to such an attack (or its natural equivalent) that as much as 95% of the U.S. population would die if such a strike were launched given our current state of readiness.

The problem, of course, is that the way back from a complete destruction of the grid and (every computer, car, plane and truck) in America is virtually nonexistent. Without power or fuels, society would grind to a halt. One possible way to help during such a crisis would be to have a series of homes that are freed from the grid. If you combine Faraday protected solar panels with battery walls (as sold by Tesla), you would have some people with power to help neighbors through the perilous times that an EMP event would cause. Being grid independent is, of course, fought by the utility companies who do not want customers who are not addicted to their services. Through the mechanism of political contributions, grid independence is frequently outlawed. Nonetheless, grid independence is an extremely desirable outcome and we should encourage this wherever possible. We should also encourage EMP resistant solar panels.

SOLAR PANELS AND MOBILITY

There is a natural symmetry between solar panels and electric cars. If electric cars protected their computer hardware with something like a Faraday cage, you could have mobility even during a total grid collapse. These are attractive features if there is a potential EMP event. However, beyond emergency help, having a fleet of cars powered by the sun would help diversify the energy market and diversity is good for the proper allocation of capital and for controlling the cost of energy.

SOLAR ENERGY WHEN IT GETS DARK

Obviously the greatest problem with solar energy is that the sun doesn't shine all the time. At the very moment you want light and power it goes to sleep. This is where batteries and energy storage devices can make such a huge difference. The storage wall concept is an attractive one because it helps to cover the greatest weakness of energy sources like solar ... which is intermittency. Solar does not run all the time, but with batteries or other energy storage devices, it can be made to be a constant source of power. In the next chapter, we talk about batteries, fly wheel devices, graphene super capacitor storage and many other load leveling options. Batteries are crucial to a modern society and we need to encourage research for better energy storage.

POSSIBLE BREAKTHROUGHS IN SOLAR TECHNOLOGY

There are a number of new approaches to solar that could lead to dramatically reduced costs. Given the typical pattern of lobbyist intervention by way of the political class, these possible breakthroughs will be fought by the existing solar industry (which likes panels made in China.) Currently, solar costs are predictable and the sales pitch is well practiced. However, disruptive technologies are always fun and here are a few ideas that could find a place in the marketplace.

1. Using perovskites to Make Silicon Panels More Efficient. Perovskites are materials with a certain type of crystalline structure that can be produced very cheaply. At Stanford, a silicon solar cell with an efficiency of 11.4% increased to 17% with perovskite. This is a significant percentage increase.
2. Thin Film Solar – The breakthrough is to use a liquid base so that solar can be sprayed on. Again perovskites are used and the potential efficiency is high. Not a real technology yet.
3. Printing Solar Panels in Rolls – Flexible solar rolls can be created using "printer" technology. Perhaps even LEDS can be made this way. Having panels that are flexible and that can be delivered in rolls has two advantages. The panels themselves potentially may be cheaper per kilowatt/hr AND there is the potential to save on installation costs as well. The EU's TREASORES Project has not reached solar panel levels of efficiency, but the work is promising.
4. Improvements In Existing Solar Panel Technology – Many people don't realize that solar panels have come a long way. Efficiencies have gone from 5% in the early years to over 20 % today. At the same time costs have come down (mostly due to economies of scale.) A 10 kilowatt system bought today sells for 1/10 of what it cost in 1990. Maybe these panels have reached their peak, but that is unlikely. The ultimate goal is to reach a cost per kilowatt level that is competitive with natural gas or coal electricity generation (counting as a "cost" the "batteries" needed to eliminate the intermittency problem.)
5. Solar Energy Trees – This is my personal favorite as it mimics processes that exist in nature. The ability to "print" trees and leaves has already been demonstrated as has the ability to harvest energy from wind and thermal changes using the same trees. My guess is that when nanotechnology takes off, so will this copycat system. Imagine a patch of trees in your yard that are as attractive as any produced by nature, yet which somehow manage to supply you with energy to power your home. Only time will tell, but the promise of nanobot trees and nanobot batteries is pretty cool to think about. The amazing thing is that these kinds of trees (and other bio mimicry "plants") would even "grow" on the Moon or Mars.

FINAL THOUGHTS ON SOLAR

We are starting to see claims that solar and wind are lower than the cost of traditional electricity. In 2014 residential rates averaged about 12.4 cents per KWH. The problems with comparing renewable costs to traditional energy costs are multiple. The first is that traditional sources of energy have been badgered by the EPA and by the states, thus raising costs artificially. In a source neutral world, "traditional" costs would be much lower. Second, the comparison is very difficult because renewables receive a lot of subsidies. They are frequently hidden and difficult to track. Third and most importantly, renewable costs do not reflect an "intermittency normalized" cost structure. The theoretical cost of wind or solar is not useful when full back-up has to be maintained because the wind doesn't blow and the sun doesn't shine on a full time basis. The part-time nature of these energy sources must be matched with something ... traditional electricity like coal or energy storage systems like flywheels or a similar device. Unless you have priced the intermittency cost into the cost of the renewable, you are just obfuscating about what these sources of energy really cost.

Having said the worst about solar, I still believe the cost of both solar and energy storage systems will drop. When they drop below breakeven (properly calculated,) the greens will be happy and I will be happy as well. While lowering CO_2 levels will slow the greening of the planet, it is important to have a diverse and vibrant energy market. Also, it is significant to note that solar is primarily a Western source of energy. The capital markets needed to advance this form of energy simply do not exist in the developing world. With 7 kilowatt hours (KWH) of energy falling on each square meter of the earth each day, this is a technology that simply can't be ignored however (for the countries that can afford it.)

Lastly, it should be re-emphasized that only consumer subsidies should be allowed. If money is given directly to corporations, *pay for play* is way too tempting. Also, all subsidies should have an expiration date. If solar can't win in the marketplace in the next decade or so, it probably won't. I know this is controversial among free market types, but keeping solar as a supported technology might allow compromise to be reached on other worthwhile energy objectives (grid hardening, flex fuels vehicles etc.)

Chapter 9 – Bring Me Batteries, More (Better) Batteries

"Innovation makes the world go round. It brings prosperity and freedom." **Robert Metcalfe**

Ultimately the secret to unlocking the vast potential of solar power is better batteries or energy storage devices. Elon Musk's Tesla Company sells a *battery wall* (about $3000) that mounts in your garage and allows you to store by day and to use by night. This means you could in theory go off-grid. The utility companies do not have your best interest at heart. They have their own interests at heart and you going "off grid" is not one of them. When you are not using their power, they are not making money. That is why they want you tied to the grid "at all times."

When I installed solar panels, I asked the rep from the utility company if I would have power during an earthquake (if the panels still worked.) "Oh no," he said. "You can get power only from the grid. When it is down, you are down." Not only that, I am limited to 90% of my annual use as to what I can sell to the grid. I am pretty sure the Borg would like this system.

What is good for the utility companies is for you to stay addicted. What is good for you and America is for the grid to become more decentralized. Why I support the solar cell / battery model is because it offers freedom ... freedom to live anywhere ... freedom to not participate in one vendor pricing.

GOING OFF GRID IS AWESOME.

If you have enough solar panels on your roof and enough storage capacity in your garage, you can be your own power company. While such an arrangement is currently expensive, it does lock in prices at today's levels and protects you from future price increases. It also allows you to run an electric vehicle with fuel provided by the sun. On a gallon of gas equivalent basis, it is like getting 100 mpg. Again, you have the potential to drop out of the transportation fuel price increase program just by having your own energy generation and storage system. Yes, there are inefficiencies and the current cost of batteries and solar panels is still marginal. HOWEVER, as time goes by, the cost of batteries will drop as will the cost of solar power. Of all the alternative forms of energy, this one has the greatest potential as it could ultimately power nearly all American homes and electric cars. Currently solar power is about ½ % of the total grid delivered energy. It will grow rapidly in the years ahead.

BATTERIES NEED TO GET BETTER

Currently we are using lithium-ion batteries to power our cars, laptops and storage walls. Economies of scale will bring down these costs, but ultimately something new is needed. Not only are there economic limitations, but lithium-ion batteries tend to catch fire. This even happened during a shipment of batteries by an air cargo plane. Catching fire is not useful. Neither is wasting money. However, by keeping and growing this part of the market, we will eventually see better energy storage systems.

NEW WAYS OF STORING ENERGY- CENTRIFUGAL or FLYWHEEL STORAGE

Yes batteries are a key component of a sophisticated and diverse energy mix. But there are other ways to store energy that should be explored. One such system relies on the old $E = 1/2MV^2$ formula (or its similar version for spinning objects). I am of course talking about centrifugal storage. Two insights have made such technologies promising. The first is that the mass should be small so that the speed can be high. The speed factor is what allows maximum energy storage because that part of the equation is squared. Doubling the mass gives you double the energy storage. BUT doubling the speed gives you speed squared extra energy storage potential. For example a heavy steel drum spinning at 1000 RPM is nowhere as efficient as say a Kevlar mass spinning at 10,000 RPM's, the smaller mass is completely swamped by the speed squared part of the equation.

The second breakthrough that has made centrifugal storage such an interesting idea is that we now have the potential to use " frictionless bearings" These are created by using magnets which repel the spinning wheel in such a way that there is nothing but air (or vacuum) between the wheel and the axel. You simply need to feed a small trickle of energy to the magnets and they can spin with virtually no loss of energy for weeks.

Well if this is such a great idea, why not power cars from spinning frictionless wheels. Actually this is a great idea until there is an accident and then having something spinning at 10,000 rpm's is suddenly a very bad idea. You have a veritable death machine shooting pieces of the wheel at very high speeds. This is akin to a shrapnel grenade ... not good ... especially in urban environments ... near schools.

However, in a buried concrete bunker with no chance of collision, such a storage system might just be viable. This is just one idea of what a different kind of battery might look like. This one has great potential. Below are a few more energy storage ideas.

ENERGY STORAGE USING SUPERCAPACITORS

Batteries store energy by way of a chemical process. Such systems have two weaknesses ... the storage and discharge is slow and the life of the battery (due to the inherent chemistry of batteries) is relatively short. Costs are beginning to come down because of the efficiencies of mass production, but these are still not optimal yet (particularly in removing the intermittency problem with certain technologies like solar.)

Supercapacitor storage technologies, on the other hand, rely on energy being stored electrostatically on the surface of a material, and do not involve chemical reactions of any kind. The charge and discharge cycles are super fast and there is virtually no limit as to how many times they can be "cycled." In terms of cycle times and life spans, the two ways of storing energy are almost mirror images of each other. HOWEVER, there are two problems with supercapacitors that chemical batteries don't have. The first is energy density. To get the same energy density, supercapacitors take up more space and weigh more. Secondly, they cost too much. In particular, the graphene storage materials (which are great because they are only one atom thick and thus lightweight) are expensive to make.

The breakthrough in supercapacitor technology is going to come from materials manufacturing. If the cost of making graphene drops, it could mean that the efficiency and energy density problems currently plaguing supercapacitors could go away. If this happens, it could be the energy storage solution that will finally make electric cars, cell phones and storage walls nearly perfect. Imagine a battery that charges in seconds instead of hours and that can be charged over and over again without loss of capacity. My guess is that significant breakthroughs will occur within the next 5-10 YEARS on this front.

ENERGY STORAGE USING FUEL CELLS

Ok so far fuel cell vehicles have lost to lithium/ion electric cars in the automotive marketplace and they have lost in a pretty big way. Honda has made a "Hydrogen" car (Clarity) for some time and its sales can be measured on a family's worth of fingers and toes (45 leases from 2008 through 2014). The Prius and other electric cars have found a niche and its pretty good size (about 120,000 plug-in units for 2014 and about 3.5 million hybrids since 2000.) But some at Toyota (the electric car king) are starting to talk about why their new hydrogen car is really going to take off.

Toyota likes the hydrogen fuel cell vehicle for two reasons: The first is cost. They believe that in the long run they will be able to make a very desirable vehicle at a reasonable price. They claim the technology is straightforward and the elimination of the expensive batteries will give the new vehicle an edge. They also like the range of their new vehicle (called the *Mirai*) It has announced that the range will be a very gas like 320 miles per fill-up.

The second reason that they like fuel cell vehicles is speed of refill. Battery charges take a long time, but hydrogen refills are much akin to filling your tank with gas. Five minutes is about the average time for a refill.

So having made the case for fuel cell cars, here is why I don't like them:

First of all, hydrogen requires a massive build out of new stations. Handling hydrogen is not at all like handling gas or methanol so this will be very expensive. Secondly as to speed, Elon Musk has demonstrated that he can "charge" the battery in his Tesla vehicles faster than you can fill your car with gas. What ... how is that possible. Simple, you pull into the station and drive over a robotic arm which will quickly "swap out" the batteries. The problem here is that this requires special infrastructure just as with hydrogen fuel cell vehicles. In the time it will take for hydrogen infrastructure to be built, I predict that a supercapacitor graphene system will emerge and end the fuel cells advantage of quick charging and range.

I have a couple of final thoughts on hydrogen. It seems that fuel cell technology is the winner between fuels cells and direct burning of H2. Not sure why this happened, but it clearly has. Second, if you look down the road a bit, there is the potential to get hydrogen very cheaply from Thorium reactors. Electrolysis of water is fairly low cost at high temperatures and that is what a LFTR reactor has in abundance (heat and electricity.) Of course with hydrogen and a bit of carbon, you could make methanol so that might be in the mix as well.

It's very exciting to see how all this will turn out. For what it's worth, I think that both vehicles will grab some sales with electric / hybrid vehicles far outselling the fuel cell vehicles in the short and medium terms. The market for hydrogen fuel cell vehicles will primarily be in California because the state has mandated extreme fleet emissions standards for CO2 and offered millions for infrastructure build out. Nothing beats zero CO2, so fuel cell vehicles may thrive in highly regulated and subsidized markets.

ENERGY STORAGE USING GRAVITY PUMPS

If you think about it, hydroelectric power is the ultimate in gravity storage. Water flows downhill and powers a generator. The water is replaced by rain water which ultimately comes from evaporation which is driven by the sun. Thus we have an excellent source of power and the perfect paradigm for gravity storage of energy.

We can do the same thing with manmade hydroelectric. We move water uphill using pumps powered by electricity and then let it run downhill (past a generator) when we need more electricity. So with a solar system, you pump by day and use by night. This is a "battery' by another name.

There are two problems with gravity storage. The first is evaporation. If you pump water into a lake, some of it will go native and travel into the sky (to become rain). Thus, a good rule of thumb is that gravity storage systems are about 90% efficient. The other problem is energy density. To store the same amount of energy as an AA battery, you need to raise 210 pounds of water 33 feet into the air. To match the energy of one gallon of gasoline you need, to raise 13 tons of water 3280 feet into the air ... not good ... not good at all.

So what we have is a fairly eloquent solution that demands LOTS of space. Thus, it would seem to have no applicability for home energy storage and limited applicability for commercial energy storage. In certain select circumstances it might be great, but as a way to solve the intermittency problem for solar (or wind) on a massive scale, it seems very limited.

ENERGY STORAGE USING COMPRESSED AIR

If you use electricity to compress air, you have something that can be uncompressed to create electricity. LightSail is a company which has attracted many high end venture capitalists, so it is considered a promising technology. However, it is a complex process because it requires specialized carbon fiber tanks and because there is a difficult to control interplay between pressure and heat. There is also the problem of safety whereby you can have catastrophic tank failures. Ultimately, safety codes may limit the amount of compression and thus raise costs beyond that of competing technologies. It reminds me a little of the problem with hydrogen burning cars because in order to achieve parity with gasoline on an energy density basis, you have to use tanks that can handle enormous pressures (up to 10000 psi.) The message here is that there may be a pressure maximum for both hydrogen vehicles and compressed air that limit efficiencies. Compressed air may have some use in an industrial setting. This remains to be seen.

ENERGY STORAGE USING HEAT PUMPS

Heat pumps are basically engines that can work in reverse. In one model, gravel is used to retain heat in one silo and cold in another during the storage phase and then to reverse the process to create energy. Heat pumps have been around for a long time and the technology is pretty well understood. Like most of these ideas, cost is the key factor. It's a faster, easier system to set up than compressed air and not as space intensive as gravity hydro. The jury is still out on this technology.

ENERGY STORAGE USING NANO BATTERIES

Nano batteries are batteries that exist at the molecular level (which offers tremendous advantages.) While these are the furthest away of all the technologies discussed, they are the most promising. Ray Kurzweil, a famous futurologist, predicts that *nano batteries* will dominate by 2030 or so and that they have the power to grow exponentially on a yearly basis. As a big Kurzweil fan, this is where I think the mega breakthrough will come from.

FINAL THOUGHTS ON ENERGY STORAGE

As you can see, there are many types of energy storage technologies being researched. It seems likely that one or more will "break through." Even the definition of what a "battery" actually is may have to be changed. If we finally get to super cheap, high density "batteries," it could change the way we think about energy in America.

Chapter 10 – Making Hay Out of a Lump of Coal

"To be or not to be ... that is the question" **William Shakespeare**

What to do with coal? Just as with Hamlet's dilemma, survival is the question. Coal stock prices have plummeted as the current administration is dead set against the industry. The EPA has promulgated anti-coal regulations that will not only raise the price for energy derived from coal, it will make it virtually impossible to produce energy from coal in the future. The U.S. is blessed with vast quantities of coal ... more than any other country on Earth. Yet we are turning our back on this important resource. This all goes back to the global warming madness and a desire by governments to increase the level of governmental control over our society. Recall that the main rule is: *energy is prosperity*. Rule 2 is that diversity of energy resources protects against falling energy stockpiles and thus falling prosperity. Simply stated we must stop the madness.

WHAT IS CLEAN COAL?

According to Wikipedia, "**Clean coal** is a concept for processes or approaches that mitigate emissions of carbon dioxide (CO_2) and other greenhouse gases" So there you have it. It's all about the CO2. But CO2 is good for the planet (see chapter 2). It makes the Earth greener, reduces desertification, and improves crop yields (up 34% just since 1990). It leads to a small, desirable increase in the temperature of the planet. According to satellite records, the planet is warming at a rate of about .11 C per decade (since 1979). Most of this is probably natural variability as just such a trend occurred in the 30's and 40's prior to the growth of manmade CO2.

In any event, since we are near the end of the current interglacial warming period (the Holocene), any warming will simply offset the solar related cooling that is very likely on the way. (Please see *Dark Winter* by John L. Casey for a thought provoking look at why the Earth will cool in the years ahead.) I note that predicted cooling may come from paucity in sunspots (happening now) or from the end of the Holocene and the start of a new ice age (which could begin anywhere from a few dozen to a few thousand years.)

Yes, we do need clean coal technologies that reduce the impact of mercury and that reduce particulate matter. BUT NO, we do not need to reduce CO2 emissions from coal. These are entirely beneficial and helpful.

COAL SHOULD PLAY A BIG ROLE IN AMERICA'S ENERGY FUTURE.

How much coal do we have? Well according to the American Coalition for Clean Coal Electricity (ACCCE), the U.S. has about 200 years of reserves at current usage levels. The U.S. has 25% of the worlds reserves (more than any other country) and has more energy in the form of coal than the entire Middle East has in the form of oil. **Energy is prosperity**. This means that a great deal of American wealth exists in the form of coal. As we transition from a fossil fuel based economy to a Thorium based economy, we should utilize every energy resource we have. The Chinese understand the *energy is prosperity* principle and they are building about one coal fired power plant per week. European counties like Germany (who foolishly abandoned nuclear energy) are also building numerous coal plants (because they have discovered that green energy doesn't scale and is far too costly and intermittent.) Japan and India and many developing countries see the welfare of their people as coming before the AGW ideology and so they are also building coal plants.

The U.S. shutting down its coal industry while the rest of the world builds coal power plants means that the net CO_2 savings will be near zero on a worldwide basis. Thus it means no net reduction in CO_2 and no reduction in worldwide temperatures. This is symbolism at its worst. Imposing economic hardship for no purpose is a pretty incredible policy. Leave it to the political class and the regulators to device a scheme that will cost hundreds of thousands of U.S. jobs, raise U.S. energy costs on consumers and businesses and reduce the greening of the planet.

For the next 25 to 100 years (depending on the emergence of transformative technologies), coal should be a cornerstone of our energy future. We have more coal than anyone else on the planet. Why would we throw this advantage away?

COAL VERSUS NATURAL GAS

Natural gas and coal are both capable of producing electricity. In the West, the fuel of choice would probably be natural gas. This is based on simple economics. They are substitutes for each other and price should determine which is used. In the East, there is an abundance of coal and that is probably the right choice for energy generation there. This assumes that the onerous regulations on coal will be lifted. If this does not happen, natural gas (if that is even permitted) will eventually replace our most abundant energy resource.

The nice thing about using coal for power generation is that it frees natural gas (by way of methanol and flex fuel vehicles) to be used as our primary source of transportation fuel. The breaking of OPEC that these technologies offer should be a crucial part of U.S energy policy and coal indirectly helps us with that goal.

If the EPA goes after natural gas as well as coal and makes it stick, I guess it's time for the song ... "Turn out the lights, the parties over."

COAL VERSUS THORIUM

Thorium reactors have the potential to create electricity cheaper than coal. (I can recommend the book *Thorium – Energy Cheaper than Coal* by Robert Hargraves. It has an excellent source of price projections.) The potential advantages are obvious however, as the energy density of Thorium is over 3 million times that of coal (per a Wikipedia energy density chart). For example, one kilogram of Thorium has the equivalent of 79,420,000 Mega Joules (MJ) of energy. Coal has 24 MJ in the same quantity of material. As Thorium reactors reach economies of scale, coal and even natural gas will fade as options.

The very fact that coal's long term replacement is starting to materialize should make any concerns about burning coal go away. Let's use it while we can.

COAL AS AN EXPORT PRODUCT

As the U.S. cuts back on the domestic use of coal (by way of frivolous regulations), there is no reason that coal should not be exported to other more enlightened parts of the world. Certainly China, Europe and most developing nations would be good export partners for this very useful product. While we would not be saving any CO_2 (a good thing), we could still pocket billions in export sales. The export of coal should be encouraged in every way possible. Given the current administration's thinking about coal, nothing less than the total cessation of coal production will be acceptable. It is imperative that a new way of looking at the world be instantiated at 1600 Pennsylvania Ave.

COAL AND THEFT ON A GRAND SCALE

Finally, I offer you one last way to think about coal. Imagine you are a community organizer kind of guy and you have the chance to destroy your enemies and transfer their wealth to your friends, would you do it? I think there is little doubt as to the answer to this question. Well guess what has happened. Coal stock prices have declined to record lows because the EPA says it is "dirty" energy. Peabody Energy, at one time the largest producer of coal, recently sold for just over a dollar a share ... down from a high of $73.73 per share when it looked like the current president might not get re-elected. So guess who just bought a million shares of Peabody and over 550,000 shares of Arch Coal? Yes, that's right ... George Soros, a mega Obama donor. Buy low / sell high. The community organizer drives down the price of coal stocks so they can be scoped up by his buddies and then perhaps we shall hear "Hallelujah," we have found a way to make coal "clean."

It's really a brilliant scheme to transfer the ownership of the coal industry from Republican leaning owners to Democrat mega donors. And hey, after office, who knows what might happen as the community organizer tries to make his way in the world by seeking funding for his library and his "charity."

Chapter 11 – Creating an American Market for Rare Earth Metals and Thorium

"Suppose you were an idiot, and suppose you were a member of Congress; but I repeat myself."
Mark Twain

The U.S. currently has one rare earth mine and they send all their output to China for processing. That means that not only telecommunication equipment, cell phones and TV's are dependent on China for key raw materials, BUT many crucial military components are also dependent on China. In any prolonged war with China, we would lose because we could not replace the key high tech components of our arsenal. This sad state has been brought to you by politicians who are strategic thinkers in the same way that Babe Ruth was a nutritional fanatic. Not only that, Thorium, which is crucial to our future, is treated as a rare earth waste material and must be encased in concrete and buried. A radical re-thinking of our policies on rare earth metals and Thorium is needed.

WHAT ARE RARE EARTH METALS?

Rare earth metals are elements that have f orbit electrons. They come mostly from the lanthanides and actinides families and they have properties that make the modern technological society possible. The rare earths have spin-up electrons which allow them to work well with other elements like iron and cobalt. These combinations allow for amazingly small magnets and other products. Headphones today are very different from the headphones of many years ago. Rare earths are the reason. The same is true for LASERS which use various rare earth elements to create different color laser lights. Modern radars and night vision goggles all rely on rare earth metals. These are just a few items that need rare earth metals to exist.

The chart below shows the lanthanides on the periodic table of elements. Just below the lanthanides are the actinides. Both rows have F orbit electrons and that is what gives these elements their special properties.

PERIODIC TABLE

RARE EARTH ELEMENTS AND THEIR USES

Below is a list of rare earth elements with their most common uses. I note that the global demand for these elements has increased substantially as new uses are constantly being found. There are plain old rare earths (21 and 61), plus light rare earths (the lanthanides) and heavy rare earths (the actinides.) The most valuable are the heavy version.

21	Scandium	Sc	Aerospace framework, high-intensity street lamps
39	Yttrium	Y	TV sets, cancer treatment drugs, enhances strength of alloys
57	Lanthanum	La	Camera lenses, battery-electrodes, hydrogen storage
58	Cerium	Ce	Catalytic converters, colored glass, steel production
59	Praseodymium	Pr	Super-strong magnets, welding goggles, lasers
60	Neodymium	Nd	Extremely strong permanent magnets, microphones, electric motors
61	Promethium	Pm	Made in the lab Only
62	Samarium	Sm	Cancer treatment, nuclear reactor control rods, X-ray lasers
63	European	Eu	Color TV screens, fluorescent glass, genetic screening tests
64	Gadolinium	Gd	Shielding in nuclear reactors, nuclear marine propulsion, alloys
65	Terbium	Tb	TV sets, fuel cells, sonar systems

66	Dysprosium	Dy	Commercial lighting, hard disk devices, transducers
67	Holmium	Ho	Lasers, glass coloring, High-strength magnets
68	Erbium	Er	Glass colorant, signal amplification for fiber optic cables, metallurgical uses
69	Thulium	Tm	Portable x-ray machines, high temperature superconductor, lasers
70	Ytterbium	Yb	Improves stainless steel, lasers, ground monitoring devices
71	Lutetium	Lu	Refining petroleum, LED light bulbs, integrated circuit manufacturing

DOES THE U.S. HAVE RARE EARTH METALS?

Yes, the U.S. has a great deal of rare earth metals. But, we have allowed China to corner the market (recently as high as 95%). China has realized that Thorium is the future of energy and they know that rare earths are central to modern manufacturing. If Thorium is found with rare earths, both are saved. If rare earths are found with iron or copper mining operations, both are saved. They do not treat anything valuable as a "waste" material.

In the U.S., there is a very different approach and Thorium is treated as "radioactive waste" by the NRC and the IAEA. The result is that special interests have swamped the general good and the U.S. is on the outside looking in when it comes to rare earth metals.

Up until about 1984, America was the largest producer of rare earth metals. China's production was essentially zero. After that point, the U.S. production steadily declined while China's steadily soared. With virtually no regulatory barriers and with the full support of the government, the Chinese became the low cost producer and eventually the dominant producer of this crucial component of modernity. The U.S. was very willing to help in this transfer of production by setting up the mirror image of the Chinese environment. We became highly regulated and non-strategic.

WHAT CAN BE DONE?

First of all, the U.S. should eliminate all regulations that impede the production of rare earth metals. These should be declared to be non-exportable because of the defense criticality of the metals and domestic markets for these crucial metals should be subsidized. National defense is too critical to leave to the vague intentions of the Chinese, who have threatened to end export of the metals because they have "their own use" for the materials. In 2014, a bill was added to the Defense Authorization Act that would allow for the U.S. to start producing its own rare earth metals by clearing away the regulatory burdens. The bill was killed not by a Senator or a Representative, but by the Defense Department who felt pressure to keep getting their REs from China (one guesses the Chinese threatened to cut them off.) WHAT? Yes, the very people responsible for our national security sided with the Chinese to keep the U.S. from changing its regulatory structure. This is the height of insanity and yet that is where we are today. To change this will require a President and a Defense Secretary that will have the courage to side with the American people. It will also require a Congress that will pass the needed legislation.

Secondly, a Thorium strategic reserve should be established so that Thorium, which is currently being treated as a waste material, can be sold to a public or private "bank" who would store the metal for future use. Once Thorium reactors begin to take off, the sale of the stored materials could create a market that would recoup much if not all of the initial storage costs. A market will eventually emerge for Thorium. We should not be burying this key element in concrete just because LFTR reactors are still a way off. With a secondary market for Thorium, the Rare Earth metals market would take off. Today, the calculation is how much can I get for the RE's less how much do I have to pay to dispose of Thorium? Financiers actually fund mines that have the lowest levels of Thorium ... not the highest levels of Rare Earths. If Thorium could be sold for a meaningful price to a Thorium Bank, the industry would be transformed.

THE CHINA PROBLEM IS EVEN WORSE THAN IT SEEMS

Imagine some innovative U.S. company comes up with a breakthrough in telecommunications, cell phones, defense products or electronics. This idea will undoubtedly involve rare earths somewhere in the process. So where will this new product be made. Will it be made in Ohio or Utah or California? Not bloody likely. Only one place on earth has the capability to deliver EVERY component of breakthrough technologies. That place is China. The products are made over there and then sold over here. U.S. jobs bleed away and are exported to China. The trade secrets for anything invented here, but made there, will bleed away as well. In effect, the Chinese navy is the fleet of cargo ships that deliver goods to America. In come the products ... out go the jobs and the capital. It is economic warfare on a scale that is unimaginable. It is not fair trade. It is asymmetric warfare and we are losing.

Want even worse news. In the 1960s the U.S. built a Thorium reactor ... a liquid fuel reactor and it ran for several years. It was ultimately killed by the military for reasons that made sense to them, but it was a very bad decision. Alvin Weinberg who created both the light water reactor and the liquid fuel reactor spent his life touting the benefits of Thorium after he was fired from his position as head of the Oak Ridge National Laboratory. He knew that as a peaceful use of nuclear energy, Thorium was by far and away the best way to generate energy. His work just faded into history. (See the appendix to Chapter 3 for a discussion of how all this came to be.) But the notes and instructions for building a Thorium reactor were left in Oakridge as PDF's. Guess who came to make copies and learn about the great American breakthrough with Thorium. Yes that's right ... the Chinese ... came to America to get our Thorium secrets and we gave it to them ... for free. Here you go. Create some new patents with these and you can dominate the new world of energy. Have a nice day.

There is a pathway forward, but it will take leadership of a kind we have not seen in a very long time.

Chapter 12 – Bringing Manufacturing Back by Making America the Low Cost Choice for Energy

"As long as the roots are not severed, all is well. And all will be well in the garden."

Chancy Gardener *(Being There)*

America is losing jobs to overseas competitors. U.S. companies (and even our own defense department) have sent jobs packing because in the short run they can make a few dollars more (cue the Ennio Morricone music.) It doesn't have to be this way and the strategies laid out in this book would be a way back for U.S. economic competiveness.

ENERGY IS THE KEY

By taking advantage of our tremendous natural resources, the U.S. has the ability to reverse the job exporting trend. Simply stated we need to become the low cost energy leader so that when manufacturers consider costs, the U.S. will look extremely attractive. *Energy is Prosperity*. That is a mantra of this book. It is also the key to bringing jobs back to America. Lower the cost of energy and the economy will boom.

DON"T EXPORT ENERGY

Well if we have all this energy, shouldn't we just flood the ports with LNG and oil and sent it abroad? After all, there is a lot of money to be had. That is what politicians who are shallow strategic thinkers and heavy imbibers of political contributions from Oil and Gas companies would have us believe. Yet do you really think that exporting energy to Asia and Europe will help bring back jobs or will it just accelerate the existing transfer? You know the answer to that. Sending energy to China or other Asian countries means we get a little extra in the energy sector, but we just continue our own doom. Think of an analogy in the developing world where they send their raw materials to developed nations which turn them into something of vastly greater economic value. We do not want to become a "developing" nation in this story.

We need to create a North American Energy Alliance (NAEA) and keep the vast bulk of our energy resources right here in North America. You want the best access to low cost energy? Then bring your manufacturing jobs to America.

BECOMING THE LOW COST LEADER

If we move ahead on all fronts, we can quickly break the strangle hold of OPEC on energy costs and the stranglehold of job movements abroad. The U.S. must think strategically. At a minimum, we must change the way we treat rare earth metals and Thorium. Even if these require subsidies, they are crucial to a vital economic future. Likewise, we must utilize natural gas for transportation fuels by allowing Flex Fuel vehicles and by authorizing the sale of methanol at the pump. We must reduce regulations on carbon based products. CO_2 is a benefit to the planet and not a liability. We must reduce corporate taxes so that the U.S. is one of the lowest cost places to do business and as a corollary we must change our tax code (which taxes income on a worldwide basis) so that profits made abroad can be brought back to America without penalty. There is nearly 2.1 trillion dollars sitting abroad and not being used to create U.S. jobs. Companies should be incentivized to bring these dollars back. This is an easy fix if the political will is there. And finally, by reducing restrictions on drilling, pipeline creation, and by providing leases for energy access to federal lands, we have the ability to dramatically change the cost of energy in America. Lower energy costs would create a bonanza in U.S. job growth. It is an economic miracle waiting to happen.

CREATING A NORTH AMERICAN ENERGY ALLIANCE

Canada and Mexico have surplus energy resources. The U.S. has vast quantities of untapped energy resources. If we would create an energy alliance and allow the free flow of energy across the borders of North America, we would create a vast zone of prosperity. A key principle of economics is that if you subsidize something you get more of it and if you tax something you get less of it. These are bedrock principles that work. So what to do with oil coming in from outside the NAEA? You tax it. What do you do for energy leaving the NAEA? You tax it. This would have the salutary effect of breaking OPEC and making the NAEA countries a place of low energy costs. Manufacturing would pour in to take advantage of our energy resources and prices.

Similarly, (although highly controversial,) we should tax imports from countries that do not allow fair trade. This would include China. Why should we pay to subsidize their growth at our expense? When the Chinese stop hacking American companies (to steal trade and military secrets) and allow for fair trade in all sectors, this tax could be stopped. Until then, a competitive equilibrium tax should be imposed.

ECONOMIC GROWTH AND AMERICAN DEBT

The U.S. currently has a national debt well on its way to 20 trillion dollars. When you add in the off budget, unfunded liabilities (probably over 100 trillion dollars) for programs such as Medicare and Social Security, you have even higher debt valuations. This is a formula for financial disaster. In fact, the U.S. Debt to GDP ratio (when you consider all forms of debt) is even higher than Greece ... a country that is by most measures bankrupt. Even just considering "official" debt only, the U.S. now has a Debt to GDP ratio in excess of 100%. We have gotten into this situation by putting huge debt loads on our children and grandchildren ... a morally dubious proposition ... fed and encouraged by morally dubious politicians.

The only way out of this financial mess is higher growth. By changing out economic growth from 2-3% to 5-6% and by adopting modest budgetary controls, the looming financial crisis can be eliminated. Having an energy policy that understands that *Energy Is Prosperity* would go a long way toward getting our financial house in order. In fact, without improved growth, there can be no happy ending to our current financial situation.

HOW LOW ENERGY COST ENERGY AFFECTS MANUFACTURING

About 25% of U.S. manufacturing is in energy intensive sectors. These include Aluminum, Chemical and Paper. In these sectors, total energy costs amount to about 7% of the COGS (on average.) In other sectors such as electronics, machinery and autos, direct energy costs may only be 2-3% of the COGS. Even *Labor* only averages about 16% of total costs. So where are the other costs? Mainly they are related to materials, design, taxes and health care. The truth is that while energy is only a part of the public policy changes needed to bring back jobs, it is a key cost item at the margin. In many key sectors, lower energy costs may even be the difference between reporting a profit or a loss. BUT to have a wider impact, there must also be a change in tax policy (to make the U.S. one of the low cost countries for taxes) AND we must change the way we handle health care. Remember the goal is to radically improve the economic growth rate and this will require a multifaceted approach.

If employer based health care were eliminated and individuals purchased health care directly (in some tax benefited way), all companies would be able to evaluate their manufacturing opportunities in a vastly simplified manner. For example, did you know there is more healthcare in the cost of a car than there is steel? That changes financial decision making in a fundamental way. While healthcare policy is beyond the scope of this book, there is no reason for healthcare to be employer based. It is a remnant of a "temporary" benefit for veterans after WWII. It has led to many misallocations of capital and has disrupted the free movement of labor (as many stay in a job just to ensure access to healthcare for a sick family member.) In any event, a combination of changes in healthcare policy, taxation, rare earth metal processing, regulations AND energy policies could lead to an American in-shoring manufacturing bonanza and transformative growth.

WHY HAVING MANUFACTURING IN AMERICA IS SO IMPORTANT

On the surface, companies move jobs to save a few dollars, but this can be a very short sighted decision. First of all, jobs moved to China, for example, come with "requirements". One of these requirements is that the patents, processes and trade secrets be revealed. That is a hit to the balance sheet of American companies that is typically not considered or discussed. Secondly, innovation is lost because process upgrades are closely linked to manufacturing. Export manufacturing, and innovation will take place abroad. This is a huge hit to productivity enhancement in the U.S. Also, the profits for say a cell phone company can be reduced substantially if only the design and sales processes take place here. The manufacturers are empowered to demand a greater share of the pie and they usually get it. This is particularly true when certain key materials (e.g. rare earths) are virtually banned in the U.S. and only available abroad.

America needs a strong manufacturing base. Only by making the right decisions on energy, taxes and other macroeconomic issues can we make this happen.

Chapter 13 – Place Restrictions on the EPA

"What light is to the eyes - what air is to the lungs - what love is to the heart, liberty is to the soul of man." **Robert Green Ingersoll**

The EPA has grown enormously since its creation in 1970 and now has over 10,000 lawyers writing new regulations and enforcing all kinds of rules. It now has the power to destroy. It can destroy an individual and take their land through fines if they deem a ditch near their house is a "wetlands." It can destroy a business if it deems you are not using the right kinds of wood in your guitar business and it can destroy whole industries (like coal) if it determines that your CO_2 emissions are "destroying" the planet. There is one thing missing in all this and that is economic sanity.

PUTTING SOME RESTRICTIONS ON THE EPA

If you do any major construction project, you will be asked to do an "environmental impact study." This is frequently used by green lawyers to stop projects. Fair enough in many cases. Nobody wants a piece of property next to a State Park to be turned into a petrochemical processing plant. On the other hand, private property rights are important and there should be a very good reason to take these away. One way to protect Americans is by putting restrictions on the EPA. In the sections below, we discuss some of the ways the EPA could be less heavy handed in their treatment of individuals and businesses. Just like the requirement that certain projects must do an environmental impact study, there must also be a requirement that the EPA must perform a financial impact study for all proposed regulations and rules (over a certain dollar amount.) This is just one of the ways that balance can be restored.

THE EPA AND ENDLESS REGULATIONS

Sadly, Congress passes laws that are frequently vague with provisos that an agency can promulgate regulations to flesh out the law. This gives enormous power to agencies and the EPA in particular. Who is really there to object if you publish regulations regarding say carbon emissions that destroy the coal industry and wipe out the capital of many generations ... not to mention job losses by the tens of thousands? Really the only check is the Supreme Court and they have been certainly willing to go along with the scheme that CO_2 is "pollution" (See *Massachusetts v. EPA* decided by a 5-4 vote in 2007).

There is an answer to the problem of unchecked power and it is fourfold in nature.

SET UP AN EPA REVIEW BOARD

First of all, there needs to be an agency to which all EPA decisions and rules can be appealed. This will require an appointment process that keeps EPA/Regulatory lawyers from dominating and empowers industry leaders instead. Perhaps a board of six could be appointed ... three by each party ... with the requirement that the appointee comes from the private sector and has no governmental experience. Add in the requirement that any ties go against the EPA and you have a very good start to stemming the power of the EPA.

CUTTING THE SIZE OF THE EPA

The EPA currently has over 17,000 people who administer the laws and regulations under its jurisdiction. Recently, with its proposed new greenhouse gas rules, it told a court that it would need 230,000 new employees and $21 billion dollars per year in additional funding. This is outrageous on every level. Carbon Dioxide is not "pollution" and lawyers are not our masters. I heard a prominent political columnist/satirist say that the difference between him and the IPCC was that the IPCC wanted to destroy western civilization and that he did not. Substitute the EPA for IPCC and you pretty much have the truth of the matter. A permanent limit of say 10,000 employees would make this extreme ambition of the EPA to destroy the American economy much more difficult to achieve.

REQUIRING THE EPA TO DO A FINANCIAL IMPACT STUDY FOR ALL PROPOSED REGULATIONS

Third, the EPA should be required to explain how their regulation is good for American business. If you want to shut down the coal industry, then your study would show billions of dollars of damage to the economy. On the other side would be what ... a possible drop in the worldwide temperature of .01 degrees centigrade. No serious person would allow such a regulation to go forward if a Financial Impact Study was required. If the rules of the review board were "ties go to plaintiffs and against the EPA", a new, properly focused EPA could be brought into being. The current system of unlimited powers is just not acceptable. The EPA has much valuable work it could be doing. Destroying the U.S. economy should not be part of it.

ALLOWING EPA LITIGANTS TO RECOVER DAMAGES

Finally, as a check on the EPA, damages should be allowed to be recovered by anyone targeted by the EPA. If the EPA is found by the Review Board or by a Court to be in the wrong, all damages to the target of the investigation should be recoverable from the EPA. This would make them think twice before they destroyed a family over some novel theory of what constitutes a "wetland" for example.

SUMMARY

In summary, we have a situation where Congress grants power to regulatory agencies because they are not willing to flesh out the law or maintain control. This empowers not only agencies, but the executive branch (who controls these agencies.) Thus, some of the most important decisions about things that affect us are taken further and further away from the people. Add in the fact that people who work at places like the EPA are never held accountable or fired and you have a formula for bureaucratic rule that is uncomfortably totalitarian leaning and certainly undemocratic. By adopting the four guidelines mentioned here, we could put a stop to the out of control EPA we currently have.

Chapter 14 – End the AGW War on Prosperity

"What has always made a hell on earth has been that man has tried to make it his heaven."

Friedrich Holderlin

The major theme of this book is that "energy is prosperity." This implies the corollary that all the steps that are being taken to reduce our energy stocks are steps being taken towards national and global poverty. What we are engaged in is nothing less than a war on prosperity. The dangers of anthropogenic global warming (AGW) are the battle cries of those that wish to effectively end the world as we know it. Only by exposing and stopping the bad science that serves as the basis for this war on prosperity can the U.S. return to its rightful path of economic wellbeing.

WHO ARE THE PEOPLE BEHIND THE AGW SCARE?

There is an interesting group of people at the IPCC and in the community that surrounds the United Nations fiefdom that has created the global warning scare. The person who headed the IPCC from 2002 through 2015 was Rajendra Pachauri. In February of 2015, Mr. Pachauri stepped down amid sexual harassment allegations. His re-distributionist views were well known.

Here are a few quotes from others closely aligned with the AGW scheme:

Quote by **Maurice Strong**, a billionaire elitist, primary power behind UN throne, and large CO_2 producer: *"Isn't the only hope for the planet that the industrialized civilizations collapse? Isn't it our responsibility to bring that about?"*

Quote by **Naomi Klein**, anti-capitalism, pro-hysteria advocate of global warming: *"So the need for another economic model is urgent, and if the climate justice movement can show that responding to climate change is the best chance for a more just economic system..."*

Quote by **Christiana Figueres**, leader of the U.N.'s Framework Convention on Climate Change: *"This is probably the most difficult task we have ever given ourselves, which is to intentionally transform the economic development model, for the first time in human history."*

Quote by **Thomas Stocker**, IPCC "scientist" and climate modeler: *"We need to devise a plan where all sectors of society contribute to the grand goal of de-carbonizing society."*

Quote from **Monika Kopacz**, atmospheric scientist: *"It is no secret that a lot of climate-change research is subject to opinion, that climate models sometimes disagree even on the signs of the future changes (e.g. drier vs. wetter future climate). The problem is, only sensational exaggeration makes the kind of story that will get politicians' — and readers' — attention. So, yes, climate scientists might exaggerate, but in today's world, this is the only way to assure any political action and thus more federal financing to reduce the scientific uncertainty."*

It is understandable that big government and one government types would like the AGW scheme. It is a way for them to realize their political and ideological dreams. For the U.N., their goal is to be the head of a world government with the ability to tax developed nations and to transfer money to poorer nations (while taking their cut). The problem is that the U.N., like the European Union, is not elected and has the potential to be veritable dictators over the world's economic and energy policies. I understand their motivations. The question is "Why don't American citizens resist this totalitarian push tooth and nail?" It completely goes against the grain of America and if it is not stopped, the world as we know it will end. In this sense, all the alarmists who say we are running out of time are right. The difference is that they think we are running out of time to end fossil fuels, while more rational people believe we are running out of time to stop the IPPC and its allies from killing economic development.

HOW COULD ONE GROUP OF IDEALOGICALLY DRIVEN PEOPLE SET THE CLIMATE AGENDA

The process here is much akin to what happened a couple of generations ago on U.S. college campuses. Dedicated people began to systematically alter what was taught, who got hired and to direct the indoctrination of students nationwide. Here is what the AGW movement has done:

1. Define the agenda – The goal of the IPCC is to "determine the role of manmade CO_2 on the climate." It could have been "determine if any human activity was causing effects that exceeded natural variability." But once you limit the discussion to just manmade CO_2, you have taken other theories completely off the table.
2. Make Sure Funding Only Goes to Pro AGW Research – If you want to get funding almost anywhere in the world, money is only available for research that is consistent with the IPCC agenda. Research funding on the warmist side exceeds funding on the skeptical side by about 3000 to 1. If you care about tenure, research dollars and academic community acceptance you will have "one point of view." It is group think and it has been bought and paid for.

3. Control the Peer Review Process – By simply taking over key journals and editorial boards, it is possible to control what gets published in the most prestigious scientific journals.
4. Control What Gets Said Online – Wikipedia and other online blogs are constantly monitored for thinking that is not part of the accepted group think. Over time the effect is substantial.
5. Sue people who exercise their free speech rights. Michael Mann sued Mark Steyn for writing a 280 word blog post which simply stated that the hockey stick graph was bogus. The suit has been going on for nearly five years with no end in sight. Even though there should be absolute freedom of speech to dispute any scientific theory, lawsuits like this have a chilling effect on opposing speech.
6. Use newspapers, magazines and TV to promote only one side of an issue. Have them tell the public that they will no longer allow skeptical points of view on their pages or channels.
7. Make claims that virtually all scientists support the AGW disaster scenario so you can marginalize your critics. The best example of this is a claim by an Australian professor that 97% of all scientific papers in the peer reviewed literature supported the view that dangerous climate change was being caused by man. A re-review of this claim showed that out of nearly 11,000 papers, less than 1/2% actually made such a claim. This falsehood is repeated again and again by the media and by the President and others. Interestingly, a peer reviewed article by Dr. David Legates and colleagues found that the methodology used by Cook et al. in the original study was seriously flawed in many ways including the failure to use the word "dangerous" (which would have been necessary to align the study with the AGW hypothesis.) The data in the original study was also not made available to the public when requested, thus making the categorizations essentially "secret." An attempt by Cook to publish in the journal *Earth System Dynamics* was flat out rejected. This 97% study was never really up to the level of serious science, yet this flawed study has been cited again and again. Joseph Goebbels, a Nazi propagandist, once said that if you repeat a lie often enough it will be accepted by the public. Apparently he was right.
8. Smear all that are skeptical of your theory by calling them names like *denier*. This is a derogative term that attempts to put one in the same category as a holocaust denier. It is offensive on every level.

WHY LYING AND BUYING SCIENCE WON'T WORK FOREVER

Because there was warning in the 1980s and 1990s, a plausible case was made by Al Gore and others that this trend would not only continue, but actually accelerate. The panic was on. Polar bears were about to become extinct. The world's major coastal cities were about to be inundated by rising seas and climate refugees by the millions were just months away. Why even the Arctic would be "ice free" by 2013. Unfortunately for the alarmists here is what has actually happened so far:

1. Polar bear populations are up from the 1950s and continue to be healthy.
2. The arctic ice is roughly where it was 10 years ago and shows no signs of vanishing.
3. Sea levels have been rising at the same modest rate since the warming that occurred after the *Little Ice Age*. No city is in any danger of inundation. Al Gore bought property at the beach.
4. Antarctica has sea ice levels at near record levels.
5. The warming that occurred in the 1980s and 1990s was virtually identical to the warming that occurred in the pre industrial 1930s and 1940s in terms of duration and slope. No warming outside of natural variability has been demonstrated.
6. According to satellite records, there has been no global warming for nearly two decades.
7. Land records published by NOAA and endorsed by NASA, show a warming trend since 2000. These records are bad on many accounts. First of all, the more recent years have been manually warmed. The raw data shows a cooling trend. The adjusted data shows a warming trend. Second, any thoughtful modification of the data would have cooled the more recent years because of the *Urban Heat Island Effect*. Third, land based records are of very poor quality because the error rate for individual temperature gauges is very high and because the land based records are not everywhere like satellite readings. Instead they are in Western countries with only limited records from Russia, South America, Africa, Asia and the Polar Regions. They do not extend to the oceans which comprise 70% of the earth's surface. Again satellites cover everywhere (including the oceans) and are not adjusted except for orbital decay patterns. So, SATTELITES ARE MORE ACCURATE (ALMOST WITHOUT QUESTION) AND THEY SHOW NO WARMING FOR NEARLY TWO DECADES.

TEMPERATURE CLAIMS VS REALITY

Claims have been repeatedly made that 2014 or 2015 or such and such a month is the warmest in recorded history. These claims are patently false. According to the more accurate satellite records, 2014 was the 7th warmest year since 1998. See the RSS dataset below:

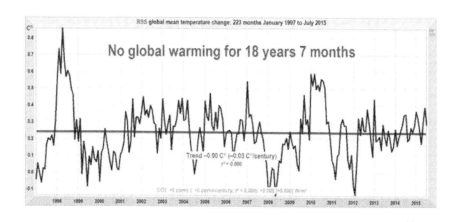

And to really put these claims in perspective, if we go back about 10,000 years (recorded history), 2014 was one of the top 3% COLDEST years during that time. See the chart below:

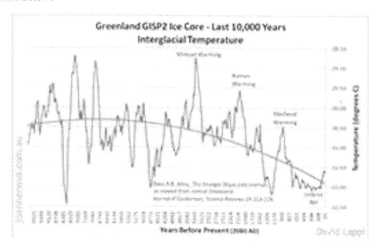

THE IPCC CLIMATE MODELS HAVE BEEN REALLY BAD

If you look at the chart shown, you will see that where we have had the ability to compare reality to the climate model projections, we see epic failures by the IPCC forecasts. Virtually every model has a higher temperature prediction (even their "low" scenario) than reality. According to the scientific method, the hypothesis has failed and must be rejected.

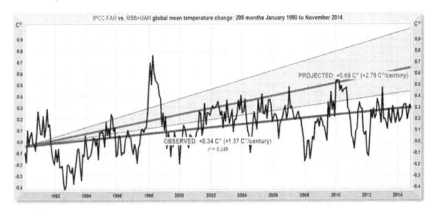

If you compare the climate models to the simple hypothesis that climate doesn't change, the simple *climate doesn't change* prediction wins. Both are wrong, of course, as climate has always changed, but it does give you perspective on the IPCC failures.

CLIMATE MODELS ARE INCONSISTENT WITH EARTH'S HISTORY

The main flaw in the climate models is that they assume positive feedback such that a little warming from doubling CO_2 will cause a water vapor feedback that will amplify the CO_2 effect by a factor of 3-9.

There is no evidence for this in Earth's history. During the Cambrian, when life exploded, CO_2 levels were over 10 times greater than today. There was no positive feedback. The planet did not burn to a crisp. The theory of positive feedback used by the IPCC just reveals how really little is known about the complex nature of climate change. More heat might well mean more water vapor, but that means more clouds and more sunlight being reflected back into space. The dynamics at play are quite complex. The history of Earth (when CO_2 was almost always higher than today) indicates that the positive feedback assumption of the models is wrong. If you believe in the scientific method, you simply must reject the dangerous AGW theory

WHAT IS MISSING FROM CLIMATE MODELS

Guess what is left out of all climate models ... the sun. That's right that big yellow object in the sky and Earth's distance from and angle to the sun are not included. *The Little Ice Age* was coincident with the *Maunder Minimum* and this in turn was caused by low solar activity. During the 1980s and 1990s, solar activity (as measured by sunspots) was high. Today solar activity is low and dropping and it looks like another *Maunder Minimum* type event is on the way.

It is important to remember that the sun is not just a source of heat, but also of radiation/solar winds. When the winds are high, cloud cover can be low and thus temperatures can rise. When solar activity is low, cloud cover can be high and temperatures may fall. Warmists frequently dismiss the sun because the temperature of the sun doesn't change much. This completely fails to take into consideration the solar winds and they don't take into consideration that small changes in the temperature of the sun can, at the margin, cause substantial changes on earth. The fact that all solar factors are excluded from climate models is very strange and most unscientific.

But it gets worse. Not only is there the complexity of sun spot activity and solar winds/cloud interactions, there is also the angle, tilt and distance of Earth to the sun. It turns out the sun's orbit relative to Earth changes due to:

1. changes in Earth's orbit around the Sun (eccentricity)
2. shifts in the tilt of Earth's axis (obliquity), and
3. the wobbling motion of Earth's axis (precession),

These three things (eccentricity, obliquity and precession) cause the Earth to go through ice age cycles (and of course, they are not in any climate models). If you look at the chart below you will see a pattern that occurs about every 100,000 years.

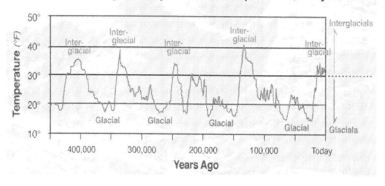

Please note that each interglacial warming period lasts about 10,000 years. The current period called the Holocene has lasted about 11,000 years and has marked the rise of man. Ask yourself, "Is it likely to get a lot warmer or a lot cooler?" It's a pretty simple answer (particularly when you understand that the theory of positive feedback is so weak.)

If these orbital pattern caused cooling events continue, we will need every drop of energy we can get. If you care about the planet and wish to avoid starvation on a mass scale, you will work to stop the AGW war against fossil fuels and other forms of desirable energy (like Thorium).

Chapter 15 – Encourage Homes and Businesses to Drop Off the Grid

"Is freedom anything else than the right to live as we wish? Nothing else." **Epictetus**

The idea of independence has a long history in America. From the New England "can do" spirit to pioneer values, the idea of self reliance is extremely American. Unfortunately, today the resistance against this is high. Nonetheless, there are many reasons to want to encourage homes and businesses to drop off the grid.

THE SOLAR PANEL / BATTERY COMBINATION OFFERS A NEW PARADUM

Yes, unsubsidized solar panels are currently uneconomic and batteries are not quite as efficient as we would like. However, both technologies will improve over time and they offer the magic of dropping off the grid. Imagine getting all your energy by day and then using it at night. In effect, you would be using personal capital to substitute for utility company capital and when the rate of return reaches an attractive level, this paradigm could be commonplace in the home/ industrial markets.

BUT WHAT ABOUT WHEN THE SUN DOESN"T SHINE?

There will be days when rain or snow reaches biblical proportions. However, if your system is rated to produce say 120% of your projected annual needs and if the energy storage system is good enough, the proverbial dark skies may not affect you. This is a design calculation. Beyond that, if you have a generator back-up that runs on say natural gas that might be enough to see you through the most statistically improbably events. There are advantages to the self contained system that could be highly beneficial. A few of these are listed below.

SAVINGS ON TRANSMISSION COSTS

For each kilowatt generated by the utility, about 6-9% is lost to transmission and distribution. This does not include costs for installing and maintaining lines meters, etc. In other parts of the world transmission and distribution losses can exceed 25%. While the U.S. savings are not super high, 6-9% is a meaningful number and would be a cost benefit component of the solar efficiency calculation.

THE ADVANTAGES OF A DECENTRALIZED GRID

Presently, carefully planned attacks on certain parts of the grid would take out vast regions of homes and businesses. Having a portion of the grid that is invulnerable to such attacks would be useful. It would mean that some members of a community would still be operating and thus able to provide assistance to neighbors if the worse happens. The same principle applies when the solar/battery home has an electric car. Such a vehicle would continue to operate even if every gas station were down because the grid was down. Electromagnetic Pulse weapons ... EMPs (discussed in the next chapter) might also be thwarted in regards to the totality of their attack if the individual homes and/or business used properly grounded Faraday boxes (unfortunately not a part of today's solar world.)

EVEN NON-EMP ATTACKS ON THE GRID CAN BE DEADLY

It should be noted that direct attacks on grid assets even with just rifles and/or explosives can produce devastating results. Most people don't realize that a sniper attack took place on the Metcalf Transmission Substation in California on April 16, 2013. The attackers fired on 17 electrical transformers which resulted in over $15 million in damage. The attack took place in the middle of the night and was quite sophisticated ... no fingerprints ... no one caught. This may have been a "trial run" and the belief is that if two or three attacks like this could be coordinated, a large part of the entire U.S. grid could have been brought down. As it was, the station near San Jose (which feeds Silicon Valley) was very close to creating massive damages. Just enough of the station was up that re-routing allowed for ongoing power to the tech capitol of America. This attack showed just how vulnerable we are even to low tech attacks.

THE ECONOMICS OF OFF GRID LIVING

Today, one would have to pay a premium to be off-grid (and in many locales this would actually be illegal.) Nonetheless the sense of freedom would be appealing to many who might be attracted to the self reliance principles at work. If you look down the road a bit, it may turn out that better solar cells or perhaps nanobot solar units might undergo dramatic price reductions in electrical energy generation costs. Similarly, nanobot storage systems or fly wheel storage systems or perhaps just improved battery systems may well become very attractive from a cost standpoint. Changing the law to allow for off grid living now would keep this option open until/if breakthroughs happen.

FREEDOM AND OFF GRID LIVING

There is a major push to "concentrate" the U.S. populations into urban centers. This has to do with the power seeking instincts of the U.N. and certain political parties. People who live in urban centers tend to vote for a more collectivist / redistributionist agenda. Agenda 21 (a U.N. idea) is being aggressively pushed by the current administration to accomplish just such a population concentration. This goes against the principles of freedom that were originally central to the fabric of America. Having the ability to live off grid goes in the opposite direction and would allow people to live anywhere. Similarly, as a principle of freedom, off grid living is a desirable goal. If technological breakthroughs happen to allow this to occur on a mass scale, the legal restrictions currently in place must not be there. If they are not removed, no amount of improved technology will allow for freedom of choice.

NANOBOT TECHNOLOGIES AND ENERGY PRODUCTION AND STORAGE

The pace of change in the world we live in is accelerating. It is the single most distinguishing feature of life in advanced civilizations today. Things that seemed impossible just a generation ago now are routine. The possibility of cyborg existence, vastly longer biological lives and virtual reality living are just decades away. Included with these miracles will be nanobot breakthroughs. Having tiny creatures with consciousness or at least functional intelligence has the potential to allow for nanobot trees and/or nanobot mini grids that could generate power in as yet un-envisioned ways (probably mimicking solutions from nature.). Similarly, having the ability to store energy using nanobots has the power to offer dramatic energy storage improvements. The point is that no one knows where the next breakthrough will come. Removing restrictions against off grid living would be a great help now because it does not restrict future possibilities.

Chapter 16 – CRITICAL: Strengthen the Electrical Grid and Infrastructure Against Electro Magnetic Pulse Strikes

"Our technological powers increase, but the side effects and potential hazards also escalate."

Alvin Toffler

The single greatest threat to America is Electromagnetic Pulse (EMP) strikes. These can occur naturally such as a major coronal ejection from the sun (a major solar storm such as this occurred in 1859) or they can occur as part of asymmetric warfare by America's enemies. The vulnerability of our electric grid which is old and getting older is beyond dangerous. It could spell the end of the country and death by starvation for most of its citizens. Nobody in Washington seems to care.

WHY ASYMMETRIC WARFARE IS SO DANGEROUS

In the good old days of U.S. / Soviet mutually assured destruction (MAD) warfare, it would have taken many weapons to wipe out the nation. Individual cities would have to be targeted and counterstrikes would be quick and highly destructive. The Soviet Union leaders did not want to die and so we had a sort of peace of the willing. Today, we live in a different world and now every violence loving nation or group (including ISIS) has the ability to wipe out America. They do not need more nuclear weapons than us, they just need two and the job is done. What wasn't known many years ago and is now common knowledge around the world is that a nuclear weapon detonation high in the atmosphere off the East Coast and another high in the atmosphere off the West Coast would create enough electromagnetic waves to wreak destruction on a colossal scale. The way these strikes work is that they destroy every computer chip in America. Unfortunately, all the technology in America uses computer chips.

ROME AND THE BARBARIANS

Many years ago, the Romans, who were the best engineers of their day, built roads all over the ancient world. These were remarkable achievements with great drainage and hard surfaces and many are still in existence today. The purpose of these roads was to allow the Roman legions to march against the barbarians and to provide a supply chain for such assaults. The road system worked as designed and Rome conquered much of the known world. Alas, these roads contained the seeds of their own destruction as the roads they built could also be used by Goths, Visigoths, Huns and Gaul's to march straight into Rome. Their greatest technological achievement was turned against them.

AMERICA AND THE ROMAN PARRALEL

Today America is a technological marvel. Food delivery systems are efficient and reliable. Highways, using cars and trucks, move people and goods all over the country. An internet allows for social connectivity on a scale never before thought possible. Every single branch of American society now uses computer chips and power. These chips are America's equivalent to the Roman roads. They are the pathway of our own destruction. Simply stated, if an EMP strike were to wipe out all the computer chips in America, virtually every power station, gas station, car, truck, computer etc. would go down with no way back up. Estimates are that 95% of Americans would die as the sudden return to an agrarian society would be devastating. People in cities would get no food, no fuel, and no services. Social order would quickly breakdown and gangs would rule for the short period of time it would take for them to starve as well. The very technology that has made possible our modern society is now our greatest weakness.

BUT WHO COULD LAUNCH SUCH A STRIKE?

The countries who could easily destroy America are China, Russia, Iran, North Korea, Pakistan, and anybody else who could acquire a couple of medium range missiles, a couple of fishing boats and two nuclear weapons. In fact, every terrorist group is a good candidate as nuclear weapons proliferate. It is worth noting that as Iran is now virtually guaranteed to acquire nuclear weapons; other Middle Eastern countries like Saudi Arabia will want their own. This nuclear proliferation in the Middle East is dangerous because of the philosophies of some of the groups there. One terrorist said, "We love death like you love life." With such people, is there any doubt that if they could destroy America, they would. They are simply not affected by MAD. Would they really care if their country became radioactive ... probably not?

WOULD WE EVEN KNOW WHO KILLED US?

The sad reality is that an EMP strike by a rogue group might not even be possible to pinpoint. Yes certain weapons have characteristics associated with their creation and we could probably identify the weapons of some countries. But what if say North Korea or Iran gave weapons and money to Al Qaeda, Hamas or some other terrorist and they use it to destroy the country we live in? Yes we might be able to strike North Korea and perhaps a few targets in Afghanistan with our submarine fleet, but where would that leave us. We would be a dying country and easy prey for countries like Russia, China just to name a few. We would be like an animal shot with an arrow ... just walking dead ... waiting for the end. ... a country with an expiration date like bad milk. The vultures would pour in and nothing would be left of the greatness than was America. That is how vulnerable we are and that is how easy it would be for us to be destroyed.

I note that the same weakness that exists for America exists elsewhere. The Israeli *Iron Dome* was made possible by computer technology. If that was destroyed, every rocket attack would be successful. Similarly Europe, which is also highly advanced, could be wiped out just as easily. Since the protection of Europe is linked to America, our destruction would almost certainly mean their own destruction as well. It would be a very different world if just two weapons were ignited in the right location.

MORE BAD NEWS: NATURE HAS AN EMP WEAPON

While the greatest risk comes from asymmetric warfare, nature also has a few possible surprises. Back in 1859, a solar flare struck Earth with a force that was very similar to what an EMP nuclear weapon strike would be like. We were not a technological nation then and other than a few telegraph stations being destroyed, no major effects were felt. Normally, the magnetosphere protects us from cosmic rays and solar flares and we get little more than really colorful *Northern Lights* when these temporarily increase. But a strike like that which occurred in 1859 would devastate America because we are extremely vulnerable. The electric companies which control the grid and the politicians and military people who are charged with defending America have shown zero interest in working to prevent this disaster. It is unforgiveable.

WHAT CAN BE DONE – NATIONALLY?

There are many things that can be done on both a grid level and a personal level to help prevent disaster. It is relatively cheap to harden the grid against solar flares and only slightly more expensive to protect the grid against EMP. The irony is that once a sufficient protection level is in place, the actual chances for an EMP strike will go down. Would someone really want to engage a partially damaged America capable of getting back on its feet quickly? Striking now while the grid is completely vulnerable is so much more appealing to enemies. Here are a few things that can be done on a national level:

1. Harden all major large transformers with devices such as the *SolidGround* system made by Emprimus. These can be installed by local electric companies

without a great deal of expense if the will is there. The will should be supplied by federal mandates as this is a crucial part of national security. Subsidies to quickly achieve the goals should be provided as well.
2. End grid tie-in which utility companies love, but are bad for individuals. Allow all users of solar energy to switch to home power only and not be required to operate exclusively as a member of the grid. Decentralization as discussed earlier would be an extremely valuable part of a national plan. Requiring solar panels to be EMP protected would also be useful.
3. Encourage food and water storage by individuals. Having a few months of supplies can make the difference between life and death in a "No Technology" world. Food storage by states, localities and FEMA could also be part of a plan to slow the impact of an EMP strike.
4. Have a sizable quantity of grid replacement components available in shielded storage. Currently replacement supplies are limited and if there were massive outages, it might take 3 years or more to secure replacements and that is an optimistic estimate. Just think about the "no way back" option if all manufacturing operations were down as a result of an EMP strike. That is why pre-positioned secure back-ups are an essential part of total grid protection.
5. We should have a comprehensive missile defense plan. This would give us our best chance to protect against low tech enemies who got a couple of nukes and missiles.
6. Develop plans at the national level for EMP recovery scenarios and EMP defense scenarios. Military leaders should be held accountable for this crucial part of our National Defense.
7. Pass legislation that requires that action be taken immediately to harden grid infrastructure. Offer incentives to utilities to comply.

WHAT CAN BE DONE – AT HOME?

There are many steps you can take to protect your home and it starts with knowledge. Here are some of the steps you can take:

1. Buy a shortwave radio and protect it in a metal Faraday cage (see info below). In the event that all communications are cut-off, it may be that the only signals you will be able to receive will come from abroad where the strike had no effect. After the initial 24 hour burst, a shortwave radio should allow you to start picking up signals from somewhere. Remember information is crucial during a disaster.
2. Protect a few LED light bulbs and LED flashlights by wrapping them in aluminum foil. This should allow you to have some light in a low electricity environment.

3. Build Faraday cages around at least one or two computers so you will have some capacity after an event.
4. Store water and food to get you through an emergency. Starvation is the greatest risk after an attack as the normal food distribution channels will be down. You should have 6 months supply of food and water at a minimum, with a year or more being even better.
5. Learn about Faraday cages as these grounding devices can save chips. Basically these are metal cages that are grounded. They can be made from wire or they can be solid (like a safe.) To work they must be grounded so that the electromagnetic pulse can be redirected away from the device you are protecting and channeled into the ground instead.
6. Have a generator and a few cans of fuel available. This may be the only source of electricity for awhile. Having lights and refrigeration will be crucial.
7. If possible keep an old car, truck or motorcycle around that does not use computer chips. In general, this means pre-1970 or so. Cars or motorcycles manufactured many years ago are not vulnerable to an EMP strike as there are no chips.
8. If you feel comfortable with weapons, it would not hurt to have a gun or two along with appropriate ammunition. Until the national grid is strengthened, you may be responsible for not only feeding you and your family, but you may be responsible for protecting them as well.

THE NEED FOR LEGISLATION IS URGENT

The local utility companies are not acting to harden the grid. They know how, but they don't want to spend money to do so. On some levels this is understandable. What we have is a national vulnerability that has put our nation at risk and the solution should be national as well. Because this is national defense problem, rules and incentives should be adopted to require hardening of grid assets and storing of replacement grid assets in secure locations. Money to accomplish these goals should be provided as well. In as few as three years, the EMP vulnerability could be greatly diminished. This would in turn reduce the likelihood of an attack as the chances of great success would disappear. The world is a dangerous enough place without the U.S. saying "here ... here is an easy way to destroy us." The time to act is now.

Chapter 17 – The Role of Fusion Power in America's Energy Future

"It sometimes seems necessary to suspend one's normal critical faculties not to find the problems of fusion overwhelming." ... **W.E. Metz**

"Fusion Power is 25 years away ... and will be for the next 50 years." So goes the joke about fusion power. The story of fusion is one of enormous promise (energy from water, no waste materials as with fission) and enormous expenditures. It has also been the story of epic failure. Creating sustainable surplus energy (whereby more energy is created than used) has been just out of reach for a long time. But perhaps this time it will happen.

WHAT IS FUSION?

Simply stated, fusion happens when two or more nuclei get close and collide. There is no conservation of matter as energy is released when the two atoms collide. Nuclei don't want to collide, so it takes a lot of persuasion (energy) to get them to cooperate. Fusion is how the sun operates, as hydrogen in the sun is constantly fused to release all the light and heat that makes life on Earth possible. This is the opposite of fission, whereby atoms are split to release energy. Fusion uses the lightest atom (hydrogen) whereas fission uses heavy atoms (like Uranium or Thorium). They are very different processes and produce energy in very different ways. Fusion creates no waste material, but fission creates many atoms with lower atomic numbers that are frequently radioactive and harmful and must be stored for hundreds of years. The simplicity of fusion is what makes it appealing ... energy from water ... with no waste. These characteristics are what have kept the fusion dream alive, despite so many years of failure.

THE PROBLEMS WITH FUSION

Massive expenditures have made this technology less than attractive. Some of the schemes for doing fusion have involved very large construction projects and have been on a scale that few but governments could afford. Certainly, the mainstream fusion projects are way beyond the scope of universities or power companies (without massive subsidies from government.) The trick is to get heat and containment to work together just right. Because the heat requirement is millions of degrees (just like the sun), there is no containments vessel made out of any known material that can hold the super heated deuterium or tritium (isotopic versions of hydrogen that come from water.) Thus the containment vessels have typically been magnetic rings or bottles. The most famous of these magnetic reactors was the Tokomak Fusion Reactor at Princeton (1982-1997.) Other type of reactors rely on using pellets that are heated so fast they don't have time to expand before the fusion reaction is started. Many other techniques have been tried, but all have met with limited success.

Another problem with fusion is that the energy doesn't really come from plain water. In actuality, isotopic versions of hydrogen must be used. Deuterium is relatively common, but tritium is much more difficult to obtain. Ultimately, tritium may have to be mined on the moon (where it is common for a variety of reasons.) Given that fuel is typically a 50/50 mix of deuterium and tritium, it is slightly more expensive than the "water as fuel" crowd would have us believe. This is NOT a deal breaker however.

The massive size and expense are really what makes this technology unappealing. If they can't be built easily and inexpensively, they just won't be able to play more than a limited role in grid energy creation. Fortunately, there are some new approaches that address these issues of size and expense.

THE LOCKHEAD SKUNK WORKS FUSION PROJECT

The Lockheed *Skunk Works* projects are legendary. The idea is to give a small team a chance to do something that very large teams are also attempting. The main funding and planning goes into the large project ... BUT the *Skunk Works* version is a "just in case" back-up plan. The amazing thing is that frequently the plan B turns out to be the winner. This happened during WWII when bombers were created by the *Skunk Works* people that were better than any other mainstream designs. This has happened repeatedly. Some of the most famous success stories were the P-38 Fighter and the U2 Reconnaissance Plane.

Recently Lockheed turned their *Skunk Works* team loose on the fusion process and early results look very promising. Their new **Compact Fusion** machine might fit in the back of a trailer or in an airline cargo hold. This makes it a vastly smaller reactor than anything previously attempted. The incredible thing is that they project a working prototype model within 5 years and a utility company ready project within 10 years. This is the most promising fusion news in a long time. In fact, other research facilities claim to be getting close to the magic breakeven point that has been so elusive. A few have actually maintained positive power generation for a few hours, so we now know that surplus fusion energy is possible.

At the core of the Lockheed breakthrough is a magnetic bottle that is made from superconducting materials. The magnetic bottle allows temperatures of tens of millions of degrees and allows continuing fusion reactions to maintain the temperatures. Heat transfer technology then converts some of the generated heat into electricity. This is very exciting research and the size, potential cost and portability advantages might make the energy produced challenge Thorium reactors for long term dominance.

FINAL THOUGHTS

If I had to guess, I believe cheap energy from fusion will not happen any time soon. The reason for this is that keeping hydrogen perpetually at 10-100 million degrees is almost beyond comprehension. In addition, if the sun is any example, we can expect lots of radiation from fusion. In fact, people travelling to Mars (without massive radiation protection), can be killed almost instantly if there is a major solar flare (radiation from fusion.) Knowing the fears that people have over traditional nuclear reactors, it seems logical that strong opposition to this would emerge as well.

There was a reluctance to include fusion in this book because of the long history of failures, but the potential is so great that even if the probability for success is low, it simply can't be ignored. Maybe, just maybe, this time Lucy will let the ball be kicked.

This chapter has been just the briefest overview of this promising but complex technology. For those who are interested in greater depth, I would encourage you to read more in the books and articles described in the Appendix to this book. It certainly is a fascinating topic.

Chapter 18 – Conservation- More Energy for Free

"A hypocrite is the kind of politician who would cut down a redwood tree, then mount the stump and make a speech for conservation." **Adlai E. Stevenson**

No discussion of energy creation would be complete without the green movements favorite place to get more energy ... using less. Conservation actually can be done in one of three ways.

The first is in theory. A study was once done that said that if everyone in America build a ten foot wall and then jumped off it at exactly the same time, you would create a tsunami that would wipe out all the coastal cities in China. Of course, if everyone in China jumped off a 3 foot wall 32.4 seconds after we jumped off our ten foot walls, then the Tsunami would be reversed and all the U.S. west coast cities would be destroyed. What we have here is national defense by theory. Conservation ideas are sometimes like this. You know ... if everyone in the Golden state stopped showering for three months and only used the toilet every third day, we could completely solve the draught problem in California. These are ideas in theory. They usually don't work in practice.

The second way to implement a conservation plan is coercion. If you are caught watering your lawn for example and you get 100 lashes for doing so, you could probably save some water. You would also not be living in any place resembling America.

Finally, you can get people to voluntarily conserve by offering incentives. If, for example, you set a policy that if you cut your water usage by 50%, your utility bill would be free, most people would change their behavior. This is the one that works because it properly takes into consideration human nature. Tax credits, free offers, penalties, fees, etc. can all be behavioral modification techniques.

I may sound like I am not much on the side of conservation, but that is not true. Conservation is great as long as it is real and as long as proper incentives are offered. There are numerous ways that energy can be saved. Here are a few:

1. Offer tax credits for solar heating
2. Offer tax credits for solar energy'
3. Offer access to the multiple passenger lanes for high efficiency vehicles
4. Offer credits for home insulation
5. Offer installing insulated windows
6. Offer credits for buying high efficiency appliances
7. Offer incentives for using LED light bulbs
8. Offer re-cycling credits for aluminum cans

All these operate off of a central theme ... you want behavior to change to save energy and you are willing to pay to make this happen. On a limited scale, such incentives are consistent with a free society. On a larger scale they are not. Thus it has even been with the scales of liberty.

Chapter 19 – The American Political System, Free Markets and Energy Policy

"It has been said that democracy is the worst form of government, except all those others that have been tried" – **Winston Churchill**

The American Political system has many strengths and many weaknesses. Foremost among our strengths is the ability to change direction by changing our elected leaders. Over time, this has served us well. Add to this, the separation of powers and the ability to create policy at the state level and you have the potential for a vibrant economic and political system. Combine strong capital markets, amazing natural resources (including abundant energy) and an entrepreneurial people and you have the most powerful country on Earth. However, our system is fraught with weakness and some of these are discussed below:

THE TRIUMPH OF THE COMMERCE CLAUSE OVER THE 10th AMENDMENT

It has been many years since the brilliance of the 10th amendment has been allowed to play out. Fearing an overly strong central government, the founders built into the constitution a provision (the 10th amendment) that limited the powers of the Federal government to *enumerated powers*. In fact, it is unlikely that the *Constitution* and the *Bill of Rights* could ever have been passed without such a provision. The 10th amendment says, "The powers not delegated to the United States by the Constitution, nor prohibited by it to the States, are reserved to the States respectively, or to the people." This was the very essence of limited government and naturally big government types felt it needed an antidote. They found it in the judicial system.

Starting in 1942, with Wickard v. Filburn, the court ruled that the *commerce clause* had wide applicability and it held that even wheat grown on a farm for its own use was "interstate commerce". It was at that point that the commerce clause became the "cure" for the limits set forth in the original founding document. It asserted that basically everything the government did was allowed because it affected interstate commerce. Funds for abortion ... no problem, funds for global warming ... no problem, funds for medical research or student loans ... no problem, shutting down a guitar factory because they used the wrong kind of woods ... no problem, destroying the coal and natural gas industries just because... no problem. No matter how far public policy strayed from the constitutionally defined "enumerated powers," it was *no problem* because the commerce clause gave everyone an out.

It's ironic that the commerce clause simply states that it is an enumerated power "to regulate Commerce with foreign Nations, and among the several States, and with the Indian Tribes". Thus it was with a novel interpretation of just four words ("among the several states") that the U.S. embarked on a new path to a socialist future. Central planners everywhere began popping corks in celebration as a new America was born.

THE INCREDIBLE RETURN ON INVESTMENT OF POLITICAL CONTRIBUTIONS

Any large business today does ROI (return on investment) analysis for all the money that the corporation invests in various projects. The after tax rate of return is compared to the after tax cost of capital, and projects are selected for funding. Virtually no project will have a return on investment as high as that of properly placed political contributions ... not the new factory, not the cloud based software platform, not the new research center. It is the political contribution that is almost always at the top of the heap. If you need to stop competition (like say methanol) or put together a monopoly or simply "earn" government contracts, money in the right hands will make these things happen.

The value of political contributions is not limited to profit seeking enterprises. Giant green organizations or unions can also assemble large dollar "donations" and "in-kind" political help to effect policy decisions. Want to kill nuclear energy or stop a dam from being built? No problem. Gather some cash, promise some votes and your wish is granted.

What we have now is a donor class that gets listened to and an electorate that gets fooled by empty promises.

LACK OF ACCOUNTABILITY

Because government workers almost literally can't be fired and because the job they have is not profit based, there is very little accountability in government. Even if you screw up big time, you will probably just be shifted around to a different department with no loss of pay.

Add to the lack of accountability the fact that government workers now make more that their private sector counterparts (salary plus pensions and benefits) and you can quickly see how the destruction of the 10th amendment has meant an ever growing government.

Big government is not where you will find entrepreneurs or new inventions. It is not where you will find fracking solutions or new battery ideas. Instead it is the place where you will find bureaucrats, deal breakers and security seekers who like very much telling actual productive people how their businesses must run. Regulatory agencies now constitute the fourth branch of government. They are unelected, and at the Federal level, they are far removed from the people. In some agencies, they are accountable to no one.

THE TURNSTILE BETWEEN ELECTED OFFICIALS, BUREAUCRATS AND LOBBYISTS

Ours was meant to be a "citizen government." Instead, we have a professional class that rules us and steals the American dream as it was originally conceived. If you lose an election ... no problem, just move to K Street and use your connections to serve the special interests. If necessary look for a job in the bureaucracy. Congress will make sure the various agencies continue to grow. There is no zero based budgeting here.

Want to see whether gambling is profitable for the casinos in Los Vegas, just walk outside and look around. Casino winnings/public losses built everything you see from Egyptian Pyramids to replicas of the Eiffel Tower. Want to see if being part of the political class pays, just check out the number of luxury car dealerships and the soaring house prices in the Washington metropolitan area (which now include parts of Maryland and Northern Virginia.) Look at all the buildings that are being erected. This has all been made possible by the massive wealth transfer between America and its Capitol. It is the Vegas trick on steroids.

IDENTITY POLITICS VS POLICY BASED POLITICS

One of the problems in American politics today is the emphasis on identity politics. If you can cobble together 90% of this group, 70% of that group and 50% of such and such other group, eventually you have a winning coalition. Policy ideas don't really matter so much except as to how they affect members of your coalition. Energy policies are far down the list of things people care about. In a world where perception matters more than reality, you can argue that you will shut down the coal and natural gas industries and the very people that will be most hurt by your ideas don't seem to care. Their votes are won through identity politics and not by way of sophisticated policy arguments.

WHAT CAN BE DONE?

Unfortunately, people are attracted to power and that is what Washington offers, so it will always be an uphill battle to constrain the growth of government. However here are a few things that might work and might be possible to pass in the right environment.

1. Appoint judges that believe in the 10th amendment. This is probably the only step that can stop the growth of government in the long run. It would require a series of three or four consecutive wins by constitutionally oriented presidential candidates. This may no longer be possible.
2. Set limits on the time people can spend in government. We already do this for the presidency and this principle should be broadly increased. Something like 12 years for Congressmen and senators would seem about right. Something like 10 years for any governmental position would also seem like a good idea.
3. Stop the crossover between elected officials and lobbyists. Put a ten year moratorium on any elected official who leaves office before he or she can hold any Federal position or be registered as a lobbyist.
4. Eliminate Federal unions. These simply perpetuate the lack of accountability problem and stifle real innovation in government.
5. Pass a constitutional amendment that limits the size and growth of government.
6. Pass a constitutional amendment to require a balanced budget.
7. Require zero based budgeting for all agencies on a regular basis.
8. Eliminate *defined benefit* pension plans. Make all plans *defined contribution* 401 K type plans. No one should be made to pay for another ones retirement (except perhaps in certain very limited cases.) The reason this is important is that this is one of the ways that government jobs are made superior to the private sector. If you are going to limit the years that someone can work in government, you will want to have a seamless pathway to move between public and private employment. The portable 401K is that pathway.
9. Restrain the EPA by adopting the proposals mentioned earlier. These include a requirement that the EPA produce a Financial Impact Study for decisions over a certain dollar amount.
10. Improve governmental accounting standards so that things like Medicare promises and Social Security commitments are properly shown as "on budget" expenses. Currently these are "off budget" balance sheet items and the true cost is hidden from the public. You would be arrested if you attempted similar schemes in the private sector.

Do I think any of these will be adopted? The odds are very low.

ENERGY POLICY AND FREE MARKETS

There are several proposals in this book that are not consistent with pure free market ideals. They are included because we don't live in the world of 1900. Instead we live in a post 10th amendment world that grants enormous powers to the Federal Government. The question is shall these be powers be used to maximize prosperity, security and liberty for the American people or will they be used to simply grow government and control the people. In each of the instances where I advocate government assistance in the energy sector, I believe it is the former that dominates.

With solar for example, advocating for limited subsidies (only to consumers and only for one additional decade) is a way of encouraging the freedom of "off grid" living (coupled with new battery technologies that are emerging.) Many might take a different opinion, but the freedom and security that a de-centralized grid offers is too attractive to pass up. Moreover, subsidies have allowed the cost of solar energy per KWH to drop by 90% since 1992.

Similarly, by offering an X-prize to build a Thorium reactor, we have a highly likely to work method to jump start a proven technology that was destroyed by governmental (military) interference. For a very modest sum, America could become the leader in this breakthrough technology and the people of this country would benefit enormously.

Likewise, adopting policies to offset the tactics of OPEC (an organization which periodically visits havoc on the U.S. energy industry by orchestrating wild price swings) would be extremely desirable. Why should we be anything less than strategic when dealing with our enemies?

In each and every case, prosperity, security and freedom are enhanced. Until the 10th amendment is restored to its rightful place and until the world becomes a less dangerous place, free markets need to be supported by tax policy that promotes American interests. Currently we offer tax breaks for breast implants and pet moving expenses. We offer tax breaks to drill for oil and to grow food for fuel. We damn well can offer tax breaks that will help us become more secure, more free and more prosperous.

Chapter 20 – SUMMARY: A Master Plan for a Better American Energy Future

"Freedom is the right to tell people what they do not want to hear." **George Orwell**

A brighter energy future is possible if the political will can be summoned. This is a big "if" as people don't usually react to "what if" as much as they react to "what is." If Americans are experiencing 2-3% economic growth, then they don't really miss the 5-6% growth that they could have had if only they had made better choices. They don't see what they don't experience. That is at the core of the problem. When "big oil" or OPEC or any number of other vested interests directly or indirectly offers money to a politician, who is on the other side? The 5-6% growth that you don't see has few supporters so they apply no political pressure. People with a keen interest in their provincial issues are powerful supporters of what they want. They know that a few dollars in donations can yield hundreds of times that in benefits. In fact, on an ROI basis, few things offer higher yields than carefully thought out K street tributes.

If flex fuel vehicles are bad for oil companies and OPEC, not only do they not get built, they wind up as "illegal." The only way to offset the K street people is to educate the American people. It is a long and tedious process, but it is the only way to create an "Other Side." So to aide in the education of what is best for America, here is a list of things that should be done:

1. Remember the overriding principle: ***Energy is Prosperity***. When we have more energy, we will be more prosperous. It's just that simple.
2. Stop the AGW war on CO_2 and fossil fuels. Fossil fuels have lifted mankind from grinding poverty, improved the water, food and air quality and have reduced climate related deaths by 98% in the last 80 years. Do we really want to return to the horrible conditions of 300 years ago before abundant cheap fossil fuels transformed humanity? CO_2 greens the planet and perhaps causes mild, desirable warming. There has been no warning for nearly two decades. The claims of the IPCC have been proven wrong again and again. Please let's honor the scientific method once more. Think first about America and demand that we return to a world of endless prosperity.
3. Offer ***X-Prize*** awards for assembly line built Thorium reactors. With the potential for tens of thousands of years of low cost energy, the whole world could be brought out of poverty. It is a CO_2 free, meltdown free, highly efficient form of nuclear reactor that offers the potential for "energy cheaper than coal." We should embrace this new technology especially in terms of the potential agreement between left and right on this issue.

4. We should reduce all restrictions on rare earth metals as they are the key to modern technology. We should stop treating the Thorium that is found with rare earths as a waste material and should start treating it as a valued asset. It should be purchased for future use and a series of thorium reserves should be setup. China's interference in the U.S. rare earth market must be stopped.
5. We should require that all new gas vehicles be *flex fuel* vehicles and we should remove all restrictions from selling methanol at the pump. With these steps we could start to use natural gas (which can be used to create methanol) as our transportation fuel. This would be a key step to breaking the undue influence of OPEC.
6. We should set up (if possible) a NORTH AMERICAN ENERGY ALLIANCE. This would allow America, Canada and Mexico to prosper from the free flow of energy across our mutual borders. It would also be an important step in bringing back U.S. manufacturing.
7. Stop the war on fracking. Fracking works, has been proven to be safe and could make a huge economic difference in many states. We should not limit the great windfall that has been presented. It is extraordinary technology and it should be fully utilized.
8. We should end all subsidies for wind energy. It is not green and it is not economic. Yes, even the green claims don't hold up when you consider how long it takes to payback the CO_2 to create and install windmills. When you consider how little useable energy you actually get out of these monstrosities given their intermittency problems, we should stop these at once. An important additional issue is the massive killing of birds. If you are truly an environmentalist, you will support the ending of this wasteful, murderous technology.
9. Keep the solar panel - battery storage model intact. This allows for freedom in a way that few things can and as these technologies get better they will help to decentralize the grid.
10. Stop the war on coal. We need energy from all sources. Over time, the fossil fuels (including coal) will fade away as Thorium and perhaps fusion take over. As a bridge fuel, it is a key component of a prosperous America.
11. Change EPA to be a more economics friendly agency by requiring a *Financial Impact Study* for all regulatory decisions over a certain amount. Allow the EPA to be sued by small claimants whose lives are often destroyed by heavy handed actions. Last, we need to set up an *EPA Review Board* with three members appointed by one party and three appointed by another. Make all decisions go against new EPA regulations if there is a tie.
12. Encourage homes and businesses to drop off the grid by removing all *grid tie-in* regulations and laws. People should be free to use their solar panels for their own use (particularly during emergencies) and not be required to provide power "only to the grid." This is a fundamental issue of freedom and fairness.

13. PLEASE PLEASE stop the EMP vulnerability that now exists. It is nothing less than the end of America if an enemy attacks this weakness. Roughly 95% of the U.S. population would die if we were hit today and it is super easy to carry out such an attack. There is a whole series of steps that need to be taken such as hardening transformers and storing replacement grid components in secure vaults. Individuals can store food and water and utilize Faraday cages to give themselves a chance at survival. No one is talking about this issue, but it is crucial to our national viability.
14. Stop funding research based on ideology. Intellectual curiosity should be enough to determine where money is spent.
15. Return science to a world of skepticism and a full on embrace of the scientific method. It is the only way forward.
16. Educate yourself on the issues and apply pressure on politicians to do what is best for America and not just what is best for the donor population. This "narrow interest gets the money and the laws" is a real weakness in our democracy. From Solyndra to bans on flex fuel vehicles, our country is made weaker by politicians who bow down to the K street \ Washington insiders. Term limits, "no lobbyist" rules for ex-politicians and a change in how donations get reported are the least steps we can take to restore "citizen democracy."

Chapter 21 – Who Are the True Environmentalists?

"The planet isn't going anywhere ... we are." **George Carlin** (*Saving the Planet*)

Mark Twain once said "The world is divided into two parts ... those that divide the world into two parts and those that do not." Nothing could be more descriptive of the great energy debate than the estimable Mr. Clemens' keen insight. Warmists and skeptics live in virtually separate universes. They talk "by" and not "with" each other and dialogue is seldom cordial. It is as if a razor has split the world into two with no way back from the side you fell on. Nothing is more surprising than the side that the environmentalists fell on. Let us take a look at a few issues of note.

THE ROLE OF CO_2 AND THE VIEW FROM THIS SIDE OF THE WALL

On my side of the great wall, there is the belief (overwhelming supported by science) that CO_2 is greening the planet. The fact that increased CO_2 causes crop yields to increase, forests to increase and deserts to shrink (because more CO_2 allows plants to use less water) is simply a function of how photosynthesis works. Estimates are that crop yields have increased (because of more CO_2) to the tune of 3.9 trillion dollars worth of extra production since 1961. The value of future increases through 2100 is estimated to be nearly 10 trillion dollars. This means many millions will not starve that otherwise would.

In addition to all these great benefits of a *Greenhouse Earth*, we also have the agricultural density issue (yields up for many crops over 300% since 1950,) which means fewer parts of the rainforest have been cut than otherwise would be and more parts of the world have returned to forest and wild areas than would otherwise be possible. All of this is because yield densities are up ... thanks in part to increased levels of CO_2.

My side supports turning the hideously ugly tar sands into wild areas and forest by extracting the energy and then replanting the entire region to make it more beautiful.

My side of the wall supports greater prosperity and more individual freedom. It supports the *scientific method*.

FROM THE OTHER SIDE OF THE WALL

And on the other side, we have support for windmills that kill 10 million birds each year. They chop them up in giant blades or beat them unconsciousness so they can die a slow and agonizing death.

But it is not just birds that die if carbon based fuels are eliminated. By moving too quickly before low cost alternative energy sources are ready, you also doom millions of people around the world to slow agonizing deaths from poverty and starvation.

In addition, a low carbon world in the developing countries is an impoverished world. Studies have shown that populations grow rapidly in poor countries and more slowly in richer ones. The economic growth needed to stop starvation and misery must come first. It is a necessity if you care about the poorest among us.

This side of the wall supports vastly larger government and diminished freedom for individuals. It also supports higher utility costs and energy shortages. It actively supports diminished economic well being here in the U.S. and around the world.

SO WHO ARE THE REAL ENVIRONMENTALISTS?

One side wants to end the slaughter of birds, see the Earth green and save the forests and grasslands from agricultural encroachment. This side wants deserts to shrink. This side wants to see the poorest among us improve their lot in life and create for all a freer and more prosperous world. This side wants to re-establish the scientific method as the basis of science. It wants to improve economies around the world because they know that prosperity leads to better environmental stewardship.

The other side wants to increase the size of an increasingly totalitarian government and to use forms of energy that are simply too intermittent to work except as supplements. They want to pretend that harming the economy will lower temperatures, when it is clear that such strategies will have no impact (run the MAGICC calculator for yourself – see Chapter Appendix for more details.) They want to replace the scientific method with "consensus." They want to kills birds by the millions and they want to transfer manufacturing and jobs to parts of the world where there are no environmental controls.

So I ask you ... which side is truly supporting the environment? And more importantly, which side are you on?

Chapter 20 - Appendix– Calculating the Impact of CO_2 Reduction Schemes

The answer that is never supplied when economy wrecking solutions are offered (like shutting down the coal and natural gas industries), is how much will these solutions lower worldwide CO_2 and worldwide temperatures. The answer is not supplied because it reduces their proposals to meaningless (but costly) symbolism. Want proof? Just run the MAGICC calculator developed by scientists at the National Center for Atmospheric Research under funding by the U.S. Environmental Protection Agency and made available on the website below:

http://www.cato.org/carbon-tax-temperature-savings-calculator

If you set everything for the maximum (all western economies, 100% reduction in CO_2, sensitivity of 4.5), you get 0.352 degrees centigrade reduction of temperatures by the year 2100. Most of the other, more realistic settings result in temperature reductions that are below the margin of error for measuring temperature. (E.g. Just the U.S., 80% reduction in CO_2, sensitivity of 1.5 yields 0.064 C by 2100; temperature margins of errors = 0.100 C)

Chapter 22 – Final Thoughts

"The greater danger for most of us lies not in setting our aim to high and falling short; but in setting our aim too low, and achieving our mark." **Michelangelo**

We live in a time of incredible change. The 20th century alone saw the rise of cars, planes and spaceships. Revolutions in medicine, telecommunication and computing changed our world in profound ways. Over 100 million people were killed in wars as various totalitarian governments raged across the planet. Yet as rapid as the change was in the 20th century, the change in the 21st century will be vastly greater … including perhaps the end of humankind as we know it.

One of the great changes will be a transformation of our energy sources. All the great concern about CO_2 emissions will soon disappear as a *Maunder Minimum* type event will descend upon the planet bringing with it lower temperatures and crop failures. Even the end of the current Holocene Inter-Glacial warming period can't be far off. Cold kills many times as many people as heat and a great cooling could soon hit America. No matter what happens (heat or cold), we will need all the energy we can get to heat or cool our homes and to bring the world and America to new levels of prosperity. *Energy is prosperity* and we need to pursue it with passion.

Over the next 100 years or so, coal and fossil fuels will start to fade away. They will be replaced by Thorium and perhaps fusion reactors and other technologies not even on the radar today. All the petty squabbling about energy sources in the current marketplace of ideas will just be an arcane historical footnote. As we settle humans (or intelligence infused robots) on Mars and build cloud cities on Venus, let us never forget that the modern world is built on energy. The change that is coming will require massive amounts of new energy. The desire to limit energy or limit certain types of energy is a call for economic and spiritual diminishment. We must resist this siren song of failure and instead align ourselves with the amazing future that is just around the corner.

Appendix - Sources and Recommendations

Here is where the intellectually curious can find out more about the topics discussed in the book. I encourage everyone to read more about these topics because my treatment was a "big picture" approach. The book was kept short for the same reason, as I wanted to cover a lot of topics with as much clarity as possible. I pride myself on the ability to make complex topics understandable. I hope that has been achieved. So for the wonderful, "filling in" part here are the places you can visit, books and articles you can read and YouTube journeys you can take. Enjoy.

Chapter 1 – Energy is Prosperity

- Book: *The Moral Case for Fossil Fuels* by Alex Epstein (This is especially good because it frames the debate in philosophical terms.)
- Book: *Heaven and Earth* by Ian Plimer
- Web: Iceagenow.com (see Co2 and prosperity)
- Web: masterresource.org (see neo Malthusian)
- Book: *Thorium Cheaper than Coal* Robert Hargraves (see chapters on third world growth and prosperity)
- YouTube – Blueprint for Western Energy Prosperity https://www.youtube.com/watch?v=qDvbt65CX_w

Chapter 2 – CO2 is Good

- Book: *Miracle Molecule, Carbon Dioxide, Gas of Life* by Paul Driessen and Ron Arnold
- Web: plantsneedco2.org
- Web: scepticalscience.com
- YouTube: Freeman Dyson (preeminent physicist) on the benefits of CO2 https://www.youtube.com/watch?v=resMHX45zPY

Chapter 3 – Thorium

- Book: *Thorium Cheaper than Coal* by Robert Hargraves
- Book: *Super Fuel: Thorium, the Green Energy Source for the Future* by Richard Martin
- Book: *What is a LFTR and How Can a Reactor Be So Safe?* By George Lerner
- Web: energyfromthorium.com
- YouTube: LFTR in Five Minutes with Kirk Sorenson https://www.youtube.com/watch?v=uK367T7h6ZY

Chapter 4 – Flex Fuel Vehicles

- Book: *Petropoly: The Collapse of America's Energy Security Paradigm* by Anne Korin and Gal Luft
- Book: *Beyond Oil and Gas: The Methanol Economy* by George A. Olah (et al)
- Web: fuelfreedom.org

Chapter 5 – OPEC

- Book: *The Price of Oil* by Julien Chartier
- Book: *The Prize: The Epic Quest for Oil, Money and Power* by Daniel Yergin (a classic)

Chapter 6 – Things that Work

- Movie: *Gasland* – This is a documentary against fracking
- Movie: *Frack Nation* – This is a documentary debunking Gasland (I liked this one better.) Taken as a set, these show how truly divided we are as a nation.
- Book: *The Boom: How Fracking Ignited the American Energy Revolution and Changed the World* by Russell Gold
- Book: *Oil Shale: Treasure Trove or Pandora's Box?* by Ronald Stites et al

Chapter 7 – Windmills

- YouTube: Bird Strike (Killing Birds) https://www.youtube.com/watch?v=CEerso.JLtRw
- Web: Wired.com (See *Unexpected downside of wind power*)
- Web: Save the Eagles http://savetheeaglesinternational.org/new/us-windfarms-kill-10-20-times-more-than-previously-thought.html
- Movie: *Windfall* (It's about the dark side of green energy)

Chapter 8 – Solar Energy

- Book: *Off the Grid: Living with Solar Energy by RD James*
- YouTube: Elon Musk Debuts the Tesla Powerwall https://www.youtube.com/watch?v=yKORsrlN-2k
- Book: *Zero to One: Notes on Start-ups or How to Build the Future* by Peter Thiel

Chapter 9- Batteries

- Web: scientificamerican.com (see energy storage articles)
- Web: Graphenea.com (Get the story of graphene super capacitors from a manufacturer.)

Chapter 10 - Coal

- Web: Soros and Coal - https://notalotofpeopleknowthat.wordpress.com/2015/08/17/soros-and-obamas-war-on-coal/

Chapter 11 – Rare Earth Metals

- Book: Rare: *The High-Stakes Race to Satisfy Our Need for the Scarcest Metals on Earth* by Keith Veronese
- Web: Which One Is the By-Product Thorium or Rare Earths? http://itheo.org/articles/Which-By-product-Thorium-Rare-Earths
- YouTube: *Thorium, Heavy Rare Earths, China & the Loss of Hi-Tech Manufacturing Jobs* https://www.youtube.com/watch?v=AkC8kItzdZI

Chapter 12 – Manufacturing

- Video: T. Boone Pickens on Energy and Manufacturing http://www.cnbc.com/id/100479643
- Web: Energy In Depth .org http://energyindepth.org/national/fracking-to-make-manufacturing-cheaper-in-the-u-s-than-in-china/

Chapter 13 – EPA

- YouTube: A Science Advisor? https://www.youtube.com/watch?v=SyK30-fl27k

Chapter 14 – The AGW War on Prosperity

- Book: *The Moral Case for Fossil Fuels* by Alex Epstein
- Book: Climate Change the Facts edited by Alan Moran (see especially the chapters on the economics of climate change)
- Web: New American http://www.thenewamerican.com/tech/energy/item/11160-sky-high-electric-bills-courtesy-of-obama-epa%E2%80%99s-war-on-coal

Chapter 15 – Dropping Off The Grid

- YouTube: The EPA's War on Off Gridders

https://www.youtube.com/watch?v=sGh3ReLRhaY

Chapter 16 – EMP Strikes

- Book: *One Second After* by William R. Forstchen
- Book: *EMP Survival: How to Prepare and Survive, When an Electromagnetic Pulse Destroys Our Power Grid* by Larry Poole

Chapter 17 – Fusion

- Book: *A Piece of the Sun: The Quest for Fusion Energy* by Daniel Clery
- Book: *Sun in a Bottle: The Strange History of Fusion and the Science of Wishful Thinking* by Charles Seife
- Web: Lockheed Martin Claims Sustainable Fusion Is Within Its Grasp http://www.eweek.com/news/lockheed-martin-claims-sustainable-fusion-is-within-its-grasp.html
- YouTube: *Is Fusion Energy Close To Becoming A Reality?* https://www.youtube.com/watch?v=2GC8lpLvrEg

Chapter 18 – Conservation

- YouTube: Saving the Planet George Carlin (very funny) https://www.youtube.com/watch?v=7W33HRc1A6c

Skeptical Websites

- Climatedepot.com
- Wattsupwiththat.com
- Hockeystchik.blogspot.com
- Iceagenow.info

Warmist Websites

- Realclimate.org
- Globalwarming–factorfictions.com

OTHER BOOKS BY ROBERT N. HENRY

Here are three other books I have written that you might enjoy (if you like Sci-Fi / Fantasy.)

Shadows of a Distant Moon (cc 2015) – A sci-fi fantasy love story featuring immortal characters who start in the 18th century and wind up on the moon in the 21st. The story explores the nature of the multi-verse, as well as the nature of genius and of good and evil. From high school drama during the depression, to war in the pacific, to alien lands, the tale ranges over many times and places. This is a super fun adventure novel for the YA or adult reader.

Mysteries of the Black Forest (cc 2015) – This is a fairy tale like fantasy about a fathers love and a daughters courage when the curse of the Norks strikes again. Set in the dark and mysterious forest of 18th century Germany, the story features an evil queen ruling over an underground kingdom of reanimated dead. It also has a time traveling dad, an improbable love story, creatures from another dimension and even a vampire or two. Targeted for a middle school audience, this is an action packed novel that could be enjoyed by readers of all ages.

Last Train to Coaltown (cc 2006) – This is a journey to the alien created virtual reality world of Elysium where many of Earth's greats have been captured and sent to the great beyond. Elysium has many "lands" including the Valley of the Moon where all men are slaves and Coaltown, a place where you are tortured by day and resurrected by night. This is a fun look at the possibilities (good and bad) for "electronic" living.